绿色城市设计丛书

拥抱蓝绿　对话城市
2020 WDS　国际联合设计作品集

EMBRACING BLUE AND GREEN, TALK TO THE CITY
2020 WORLD DESIGN STUDIO COLLECTION

主　编 | 高　翅
副主编 | 王　玏　章　莉　张　炜　邵继中
Editor in Chief | GAO Chi
Deputy Editors in Chief | WANG Le　ZHANG Li　ZHANG Wei　SHAO Jizhong

中国建筑工业出版社

图书在版编目（CIP）数据

拥抱蓝绿　对话城市：2020WDS国际联合设计作品集=EMBRACING BLUE AND GREEN，TALK TO THE CITY：2020 WORLD DESIGN STUDIO COLLECTION / 高翅主编；王玏等副主编. —北京：中国建筑工业出版社，2022.3

（绿色城市设计丛书）

ISBN 978-7-112-26752-1

Ⅰ.①拥…　Ⅱ.①高…②王…　Ⅲ.①城市规划—建筑设计—作品集—世界—现代　Ⅳ.①TU984.2

中国版本图书馆CIP数据核字（2021）第211273号

责任编辑：郑淮兵　陈小娟　王晓迪
责任校对：王　烨

绿色城市设计丛书
拥抱蓝绿　对话城市
2020 WDS　国际联合设计作品集
EMBRACING BLUE AND GREEN, TALK TO THE CITY
2020 WORLD DESIGN STUDIO COLLECTION
主　编 | 高　翅
副主编 | 王　玏　章　莉　张　炜　邵继中
*
中国建筑工业出版社出版、发行（北京海淀三里河路9号）
各地新华书店、建筑书店经销
北京雅盈中佳图文设计公司制版
北京富诚彩色印刷有限公司印刷
*
开本：880毫米×1230毫米　1/16　印张：13½　字数：452千字
2022年3月第一版　2022年3月第一次印刷
定价：186.00元
ISBN 978-7-112-26752-1
（38553）

前言

“全球视野、本土行动、特色发展”，是华中农业大学风景园林坚持多年的办学理念。从造园教育初创时引进留学教师，到后来不断派送教师出国访学、学生出国交流，逐渐发展出暑期联合设计工作坊、海外实习、国际学术论坛、国际竞赛、国际联合培养等多样方式，“走出去、请进来”，不断增强国际合作交流。自 2015 年始，由华中农业大学、美国克莱姆森大学（Clemson University）和埃及艾因·夏姆斯大学（Ain Shams University）合作开设 WDS（world design studio）联合设计课程，三所学校每年轮流执掌教学主场，提供统一选题，师生校际交叉组合，组织实施课程教学。数年来，这些举措在帮助学生了解学术动态、开拓国际视野、培养国际交流与合作能力、加强跨文化交流诸方面，起到积极推动作用。

2020 年的 WDS 联合设计课程以“武汉长江新城谌家矶片区城市设计”为主题。长江经济带横跨我国东、中、西三大区域，对中国战略发展全局具有重大意义。谌家矶片区作为汉口的重要组成部分，自开埠以来，就承担了对外通商联络的港口功能，造就了汉口开放、共享、包容、创新的特性。本次课程采取“项目制”教学方法，以问题为导向，激发学生主动思考场问题，在主动参与、团队协作过程中思考滨水新城的绿色城市设计途径，提出蓝绿城市空间体系构建、无人驾驶绿色城市构建、有机融合多元城市功能等方法与策略，营建人与自然和谐共生的城市家园。课程采取线上教学模式，运用“Zoom 视频会议 +MOOC+QQ 群”进行“设定任务 + 分组汇报 + 指导答疑”的闭环教学，校际联系采用“邮箱 + 微信 + 互换 PPT”的沟通形式，构建“教学做一体化”的课堂。

风景园林旨在创造持续美好生活，风景园林教学为创造持续美好生活培养创新人才。风景园林国际联合设计课程的开展，基于全球视野，审视并关注人类对美好生活的向往，在国际交流与合作中不断碰撞出思想火花，凝聚成文化交融的结晶。现阶段的设计课程教学成果虽只是“芽苞初放”，我们相信，在不久的将来定会有“绿意盎然”。

华中农业大学园艺林学学院副院长

张斌

Foreword

"Global vision, local action, and characteristic development" are the school-running philosophy that the Department of Landscape Architecture of Huazhong Agricultural University has adhered to for many years. From the beginning of the garden construction education, we hired overseas teachers, and then continued to send teachers to study abroad, students to exchange abroad, and gradually developed summer joint design workshops, overseas internships, international academic forums, international competitions, and international joint training. Following the way of communication of "go out and invite to come in",we continuously strengthen international cooperation and exchanges. Since 2015, Huazhong Agricultural University, Clemson University, and Ain Shams University in Egypt have jointly opened WDS（world design studio）joint design courses. The three schools take turns to take charge of the teaching arena and provide unified topic selection every year. The inter-school cooperation and exchanges between teachers and students jointly organize and implement curriculum teaching. In the past few years, these measures have played a positive role in helping students understand academic trends, broadening their international horizons, cultivating the ability of international exchanges and cooperation, and strengthening cross-cultural exchanges.

The theme of the WDS joint design course in 2020 is "Urban Design of Chenjiaji Area of Wuhan Yangtze River New Town". Spanning the eastern, central and western regions of China, the Yangtze River Economic Belt is of great significance to China's overall strategic development. As an important part of Hankou, the Chenjiaji area has undertaken the port function of foreign trade and liaison since its opening as a port, creating Hankou's characteristics of openness, sharing, inclusiveness and innovation. Adopting the method of "project-based" teaching, this course is problem-oriented and stimulates students to consider problems of the site actively. We aim to inspire students to think about the green urban design approach of the Waterfront New Town in the process of active participation and teamwork, propose blue-green urban space system, driverless green city construction and organic integration of multiple urban functions and other methods and strategies to build urban homes where man and nature live in harmony. The course adopts the online closed-loop teaching mode of "Zoom video conference + MOOC + QQ group" for "task setting + group report + guidance and answering questions". The inter-school contact adopts the communication form of "email + WeChat + exchange of PPT" to build an integrated classroom of "teaching, learning and doing".

Landscape architecture aims to create a sustainable and beautiful life, and landscape architecture teaching cultivates innovative talents for creating that life. The establishment of the international joint design course of landscape architecture is based on the global vision and the scrutiny and attention to human yearning for a better life. It constantly collides with thought sparks in international exchanges and cooperation and condenses into the crystallization of cultural fusion. Although the results of the design course teaching at this stage are only "buds", we believe that there will be "greenery" in the near future.

Deputy Dean of College of Horticulture and Forestry, Huazhong Agricultural University

ZHANG Bin

国际联合设计工作坊（WDS）——美国

克莱姆森大学

风景园林

建筑学院

WDS 参与院校

国际联合设计工作坊（WDS）由埃及艾因・夏姆斯大学工程学院建筑系、美国克莱姆森大学风景园林系以及中国华中农业大学风景园林系合作创办。

WDS 职责

WDS 为上述三所大学提供了一个平台，在参与院校的使命、目标、战略和意向下，各院校的教授、研究人员和学生通过多学科国际合作资助的研究、咨询、社区参与以及跨文化丰富课程、教学方法和学习成果，得以进行知识交流。

WDS 历史沿革

自 2006 年以来，克莱姆森大学风景园林系的师生便与艾因· 夏姆斯大学建筑系的师生一同在跨文化联合设计工作坊中工作，并着眼于世界上一些最为重要遗址的社区参与和设计。两所大学共同签订合作谅解备忘录并建立学术伙伴关系，因而取得了多个同国际、区域或是国家合作的社区参与项目。在过去十年里，各工作室获得了极高水准的表彰及奖项，同时也在国内或国际上得到了认可。包括麻省理工学院阿迦汗项目，国家地理学会，基石索诺玛，以及卢克索省长，埃及总理，美国风景园林师协会（三州奖）、美国风景园林师协会 - 南卡罗莱纳奖、风景园林学教育委员会（CELA）国家和国际奖，艾因・夏姆斯大学卓越奖。2016 年，华中农业大学风景园林系加入国际联合设计工作坊，成为第三个合作伙伴。2017 年三所大学共同参与了中国武汉四新新区的发展规划；2018 年 3 月应埃及红海总督的邀请，为实现赫尔格达市的社区经济和旅游发展进行规划设计；2020 年世界城市发展大会讨论了中国武汉市的城市设计问题，克莱姆森大学专注于“新城市生态农业健康城市”这一设计方案。2020 年初新冠肺炎疫情暴发，虽然老师和学生无法当面交流，但三方仍通过线上平台成功完成了整个设计。

WDS 目标

国际联合设计工作坊的学生来自非洲、北美、亚洲和欧洲四大洲，是第一个以三所大学的学术资源培养学生在建筑与风景园林专业的相关技能的工作坊。意在促进多学科间的交叉融合，包括工程相关专业、生物、自然科学、农业、艺术以及人文学科；旨在提供基于项目的课程，这些项目将广泛关注当下世界各地的社区参与、经济发展、创新、可持续性、弹性、先进的材料及制造技术、大数据及分析、能源及交通等。

WDS 影响

国际联合设计工作坊在全球社区参与、环境可持续性、弹性和创新等方面的经历得到了极高的关注。经常参与一些备受重视的项目，并赢得了大使馆、州长、部长、大学校长、市长和公共机构的赞誉。目前已经斩获国家和国际教学奖以及学生工作奖，师生合作完成的研究发表在国内外期刊和会议上，参与合作的教师和学生也被邀请参加著名的学术会议和研讨会。

自国际联合设计工作坊开展合作项目以来，已有超过 150 名美国学生、300 名埃及学生和 150 名中国学生参加了 WDS 合作项目，其中 75 名学生跟进了近期项目。学生们在合作中成为挚友，一直到工作后也保持着联系。此外， WDS 的媒体报道、大学研讨会、公共研讨会也向更多的学生、校友和大学介绍了国际教育和全球参与在解决重大全球问题中的作用，以及文化意识和文化理解在这一过程中的重要性。

Hala F. Nassar 教授， 美国风景园林教育委员会

风景园林教授

建筑学院

克莱姆森大学

前副主席

美国风景园林教育委员会

地址 3-104 Lee Hall | Clemson, SC 29634

电话（864）656-2499

World Design Studio（WDS）—USA

Clemson University

Landscape Architecture

School of Architecture

WDS Description & Participating Universities

The World Design Studio（WDS）incorporates a collaboration between the Department of Architecture, Faculty of Engineering, Ain Shams University（ASU）, Landscape Architecture in Clemson University in the United States of America and the Department of Landscape Architecture in Huazhong Agricultural University（HAU）in China.

WDS Mission

The WDS provides a platform for: knowledge exchange between the collaborating Universities, their professors, research faculty, and students through multidisciplinary international collaborative funded research, consultation, community engagement, and cross-cultural enrichment of curriculum, pedagogy, and learning outcomes in the context of the participating schools and universities' missions, goals, strategies, and objectives.

WDS Program History

Since 2006, Clemson University Landscape Architecture faculty and students have been working with Ain Shams University Architecture faculty and students in parallel collaborative cross-cultural design studios focusing on community engagement and design for some of the most important heritage sites in the world. This academic partnership, celebrated by an MOU between the two universities, has yielded multiple international, regional and state collaborative community engagement projects. The design studios have achieved the highest level of recognition, review and awards and have been recognized in national and international forums during the last decade, including the Aga Khan Program at MIT, National Geographic Society, Cornerstone Sonoma, and by the Governor of Luxor, the Prime Minister of Egypt, the American Society of Landscape Architects（Tri-State Award）, the American Society of Landscape Architects - South Carolina, the Council of Educators in Landscape Architecture（CELA）national & international awards, and Ain Shams University Awards of Excellence. In 2016, Huazhong Agricultural University（HAU）Department of Landscape Architecture became the third partner by joining the International Collaborative Design Studio. Three partner universities worked on community development design of the Sixin District in Wuhan China in 2017. In March 2018, the three studio were invited by the Governor of the Red Sea in Egypt to address community economic and tourism development goals for the city of Hurghada. In 2020, the WDS addressed Urban Design of the city of Wuhan in China. Clemson University Studio focused on a design proposal for a "New Urban Ecological Agricultural Healthy City". While the outbreak of Covid-19 pandemic in early 2020 that prohibited face-to-face meeting of the studio faculty and students, the three studios were able to finish the design process successfully using online and virtual platforms.

WDS Objectives

The WDS design studio includes students from four continents, Africa, North America, Asia and Europe. It is the first studio of its kind, employing the knowledge and expertise of the three Universities in developing student skills in

Architecture and Landscape Architecture. The WDS is intended to grow to incorporate a wider range of disciplines, including: a range of engineering specializations, the biological, natural sciences and agriculture, the arts and humanities. It is intended to offer project-based coursework broadly focused on contemporary community engagement, economic development, innovation, sustainability, resiliency, advanced manufacturing and materials, analytics and big data, energy and transportation, potentially anywhere in the world. The WDS includes undergraduates and post graduate students working on their Master's or PhD degrees.

WDS Impact

The World Design Studio's record of global community engagement, concern for environmental sustainability, resiliency, and for innovation has garnered attention at the highest levels. The studio regularly engages in high-profile projects, earning accolades from embassies, governors, ministers, university presidents, mayors and public agencies. WDS projects have been recognized through national and international awards for teaching, and for student work. Collaborative research between faculty and students is published in peer-reviewed journals and conferences nationally and internationally. Collaborating faculty and students have received invitations to prestigious academic and professional conferences and workshops.

Additionally, since the beginning of the World Design Studio partnerships, more than 150 American, 300 Egyptian, and 150 Chinese students have participated in WDS collaborations, with 75 students participating during the most recent offering. Students involved in these collaborations develop lasting friendships that have continued into their professional lives. Peripherally, WDS media coverage, college symposia, public workshops and reviews have introduced extended groups of students, alumni, and university communities to the role of international education and global engagement in addressing significant global issues, and the importance of cultural awareness and understanding in this process.

Hala F. Nassar Ph.D., FCELA

Professor Landscape Architecture

School of Architecture

Clemson University

Past Second Vice President

Council of Educators in Landscape Architecture

3-104 Lee Hall | Clemson, SC 29634

Tel（864）656-2499

国际联合设计工作坊——埃及

艾因·夏姆斯大学正在采取具体措施促进学院和研究项目的国际合作。国际联合设计工作坊 (WDS) 是我们大学教育的非凡经验之一，我们为其多年来的发展感到非常自豪，也为来自三个大洲的三个合作伙伴间（埃及艾因·夏姆斯大学工程学院建筑系，美国克莱姆森大学风景园林系以及中国华中农业大学风景园林系）的合作感到非常高兴。

埃及高等教育和科学研究部一直致力于支持高等教育的国际化，并将埃及定位为国际学生的中心。得益于埃及高等教育和科学研究部的努力，WDS 被视作一个独特的设计工作坊，为促成更多国际合作奠定基础。这一多学科、跨文化的设计工作坊以三所大学的学术资源及各类设施，培养来自非洲、北美、亚洲学生们的专业技能。

这个跨文化联合设计工作坊始于 2007 年，由埃及艾因·夏姆斯大学工程学院的建筑系和美国克莱姆森大学建筑学院的风景园林系合作建立。两所大学共同签订合作谅解备忘录并建立学术伙伴关系，因而取得了多个同国际、区域或是国家合作的社区参与项目。在过去十年里，设计工作坊获得了极高水准的表彰及奖项，同时也在国内或国际上得到了认可。包括麻省理工学院阿迦汗项目，国家地理学会，基石索诺玛以及卢克索省长，埃及总理，美国风景园林师协会（三州奖）、美国风景园林师协会 - 南卡罗莱纳奖、风景园林学教育委员会（CELA）国家和国际奖，艾因·夏姆斯大学卓越奖。

2016 年，华中农业大学风景园林系加入国际联合设计工作坊，成为第三个合作伙伴。这三所大学合作参与了许多社区发展设计，如 2017 年中国武汉四新新区，2018 年赫尔格达市海洋区域和人行道的规划。2019 年，国际联合设计工作坊谅解备忘录在美国克莱姆森正式签署。在今年，国际联合设计工作坊项目是针对查尔斯顿市海平面上升威胁下的城市和建筑解决方案。2020 年，在国际联合设计工作坊的项目中，埃及方面应对全球传播的新冠肺炎的持续挑战，期间以武汉市延伸区项目选址为例，从中开发应用创新的设计理念，以构建在城市和建筑尺度上可采用健康可持续方式的疫情下的弹性建筑环境。

World Design Studio—Egypt

Ain Shams University is taking concrete steps for fostering international collaborations in its faculties and study programs. One of the remarkable experiences fostered by Ain Shams University is the World Design Studio (WDS) which we are very proud of how it has developed over the years. We are very glad with this collaboration that is taking place between the three partners from three continents; Faculty of Engineering at Ain Shams University in Egypt, the Landscape Architecture Program, School of Architecture at Clemson University in the United States, and the Department of Landscape Architecture at Huazhong Agricultural University in China.

The Egyptian Ministry of Higher Education and Scientific Research is always keen on supporting the internationalization of Higher Education and positioning Egypt as a hub for international students. Where, WDS is considered as a unique design studio that shows the Ministry's effort to pave the way for such international collaborations to take place and thrive. The multi-disciplinary and cross-cultural design studio employs knowledge, expertise and diverse facilities of the three universities in developing the skills of students from Africa, North America and Asia.

This collaborative cross-cultural design studio has started Since 2007, between the Department of Architectural Engineering, Faculty of Engineering at Ain Shams University in Egypt, and the Landscape Architecture Program, School of Architecture at Clemson University in the United States. This academic partnership, celebrated by an MOU between the two universities, has yielded multiple international, regional and state collaborative community engagement projects. The design studios have achieved the highest level of recognition, review and awards and have been recognized in national and international forums during the last decade, including the Aga Khan Program at MIT, National Geographic Society, Cornerstone Sonoma, and by the Governor of Luxor, the Prime Minister of Egypt, the American Society of Landscape Architects（Tri-State Award）, the American Society of Landscape Architects - South Carolina, the Council of Educators in Landscape Architecture（CELA）national & international awards, and Ain Shams University Awards of Excellence.

In 2016, the Department of Landscape Architecture at Huazhong Agricultural University in China became the third partner by joining the International Collaborative Design Studio（ICDS）. The three partner universities worked on many community development designs, such as: the Sixin District in Wuhan China in 2017, the development of the marine area and walkway in the city of Hurghada in 2018. In 2019, the World Design Studio MOU was officially signed in Clemson, SC in USA. Also, in this year, the WDS project was an urban and architecture solutions concerning Charleston City's sea level rise challenge. In 2020, WDS Egypt tackled the ongoing challenge of the globally spread COVID-19, taking the project location at the extension of Wuhan city as a case study for developing and applying innovative design ideas for a pandemic-resilient built environment that can function in a healthy and sustainable way on both the urban and building scales.

课程介绍

1. 课程简介

城市的发展是动态变化的层积过程，各种要素的关联交织，形成了城市自然系统与人工系统的融合发展。基于城市特定片区发展所面对的复杂要求，如何找准特色突破方向，同时协调与处理城市多种矛盾问题，是绿色城市设计面临的重要任务之一。设计选题结合当前国土空间规划中的热点问题，与生态学、社会学、城乡规划、建筑学和人类学等学科内容进行交叉研究，突出设计的复杂性和综合性。课程旨在培养和加强学生综合分析问题、独立自主思考、团队沟通合作等方面的能力。

2. 基址概况

武汉长江新城起步区规划范围东至武湖泵站河，南至长江北岸，西至滠水河、府河，西南至张公堤路，北至四环线，面积约 58.8km^2。起步区可划分为总部基地服务片区，青年活力乐居片区、国际交往服务片区和产城融合创新片区，其主要功能定位为区级商务办公中心、区级商业、区级公共服务中心、居住和城市公园。

3. 设计要求

结合城市新区发展的目标和定位，通过风景园林学的途径，创造一个健康、活力、优美、舒适的新城片区，要求设计立意新颖，设计手法创新，设计内容满足相关规范标准。

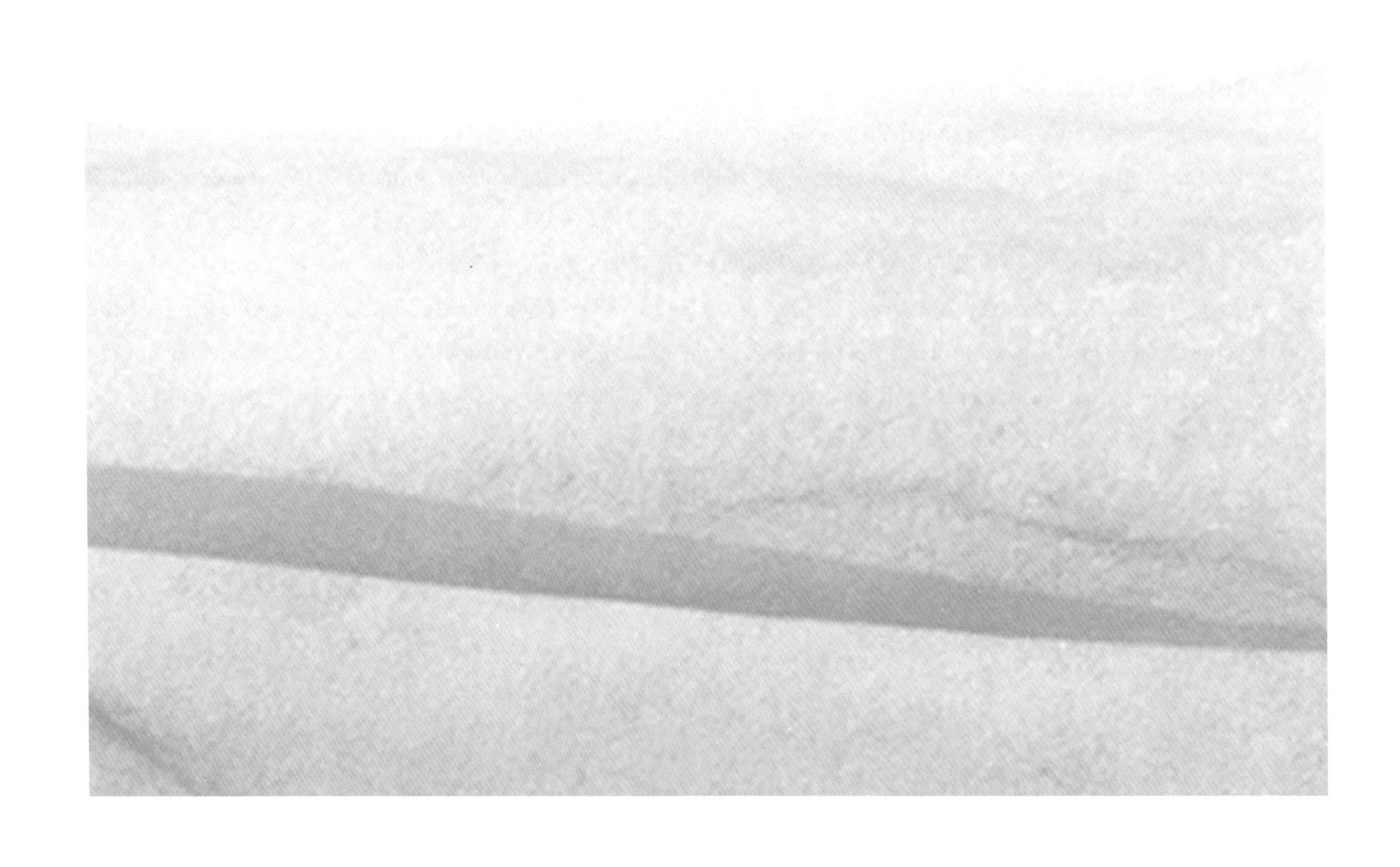

Course Introduction

1. Course introduction

The development of a city dynamic process of stratification. The interweaving and interrelation of various elements has formed the integrated development of urban natural system and artificial system. Based on the complex requirements faced by the development of specific areas of the city, how to find the direction of characteristic breakthroughs, and at the same time coordinate and deal with multiple contradictions in the city, is one of the very important tasks for green urban design. The design topic is combined with current hotspots in territorial spatial planning, and an interdisciplinary approach between ecology, sociology, regional planning, architecture and anthropology as well as other disciplines, highlighting the complexity and comprehensiveness of the design process. The course is aimed at cultivating and strengthening students' abilities in comprehensive analysis of problems, independent thinking, and team cooperation.

2. Site condition overview

The planning starting area of Wuhan Yangtze River New City extends from the Wuhu Pumping Station River in the east, the north bank of the Yangtze River in the south, the Shishui River and the Fuhe River in the west, Zhanggongdi Road in the southwest, and the Fourth Ring Road in the north, covering an area of about 58.8 square kilometers. The starting area can be divided into the headquarters base service district, the youth vitality happy home district, the international exchange service district and the industry-city integration innovation district. The main functions of the area includes the district-level business, commerce, and public service sectors, residential units and urban parks.

3. Design requirements

The planning and design aim to achieve the goals of the new city development, and to create a healthy, vigorous, beautiful and comfortable new city with the approach of landscape architecture. The outcome requires novel design ideas, innovative techniques, and contents that meets relevant norms and standards.

WDS 指导老师 WDS Instructors

PROF. DR.
SHAO Jizhong

ASSOC. PROF.
DR. WANG Le

LECT. DR.
ZHANG Li

ASSOC. PROF.
DR. ZHANG Wei

PROF. DR.
ROBERT HEWITT

PROF. DR.
HALA NASSAR

PROF. DR.
MOHAMED
AYMAN ASHOUR

PROF. DR.
YASSE MOHAMED
MANSOUR

PROF. DR.
MOSTAFA
REFAAT AHMED

ASSOC. PROF. DR.
ASHRAF ABD EL
MOHSEN

ASST. PROF. DR.
AYMAN AHMED
FARID

ASST. PROF. DR.
AHMED KHALED
ABD EL-HAMEED

PROF. DR.
GHADA FAROUK

PROF. DR.
SHAIMAA KAMEL

PROF. DR.
MARWA KHALIFA

ASST. LECT.
MAI ADEL FATHY

ASST. LECT.
NADA TAREK
ESMAEL

TEACHING
ASST. MOAMEN
NASSER

TEACHING ASST.
AMR AHMED
SALAH

TEACHING
ASST. MOSTAFA
KHALED

CONTENTS / 目录

基址分析 Site analysis
上位解读 Preliminary Plan Interpretation

项目规划目标：长江新城被定义为“未来之城”，是武汉“1+5+1”城市发展目标体系的一部分，目标是打造代表城市发展最高成就的展示区、全球未来城市的样板区。

The planning goal: Yangtze River New Town is defined as the “City of the Future” and is one of Wuhan’s “1+5+1” urban development target system. The goal is to create an exhibition area that represents the highest achievements in urban development and a model area for future cities in the world.

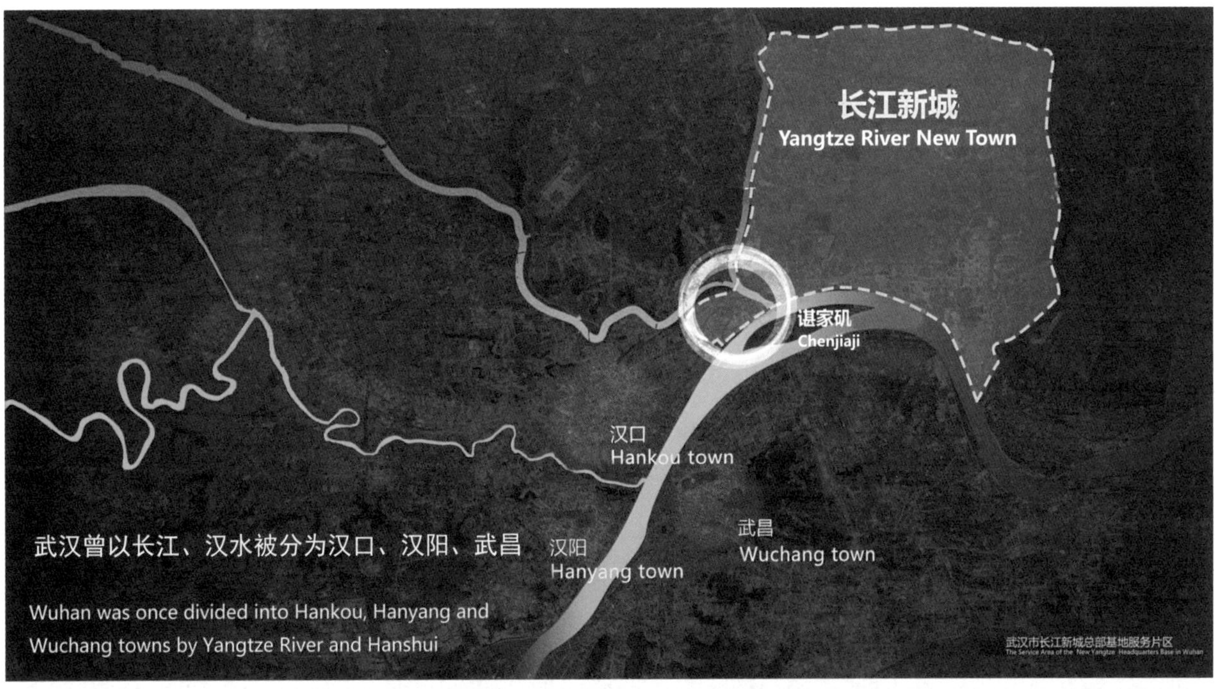

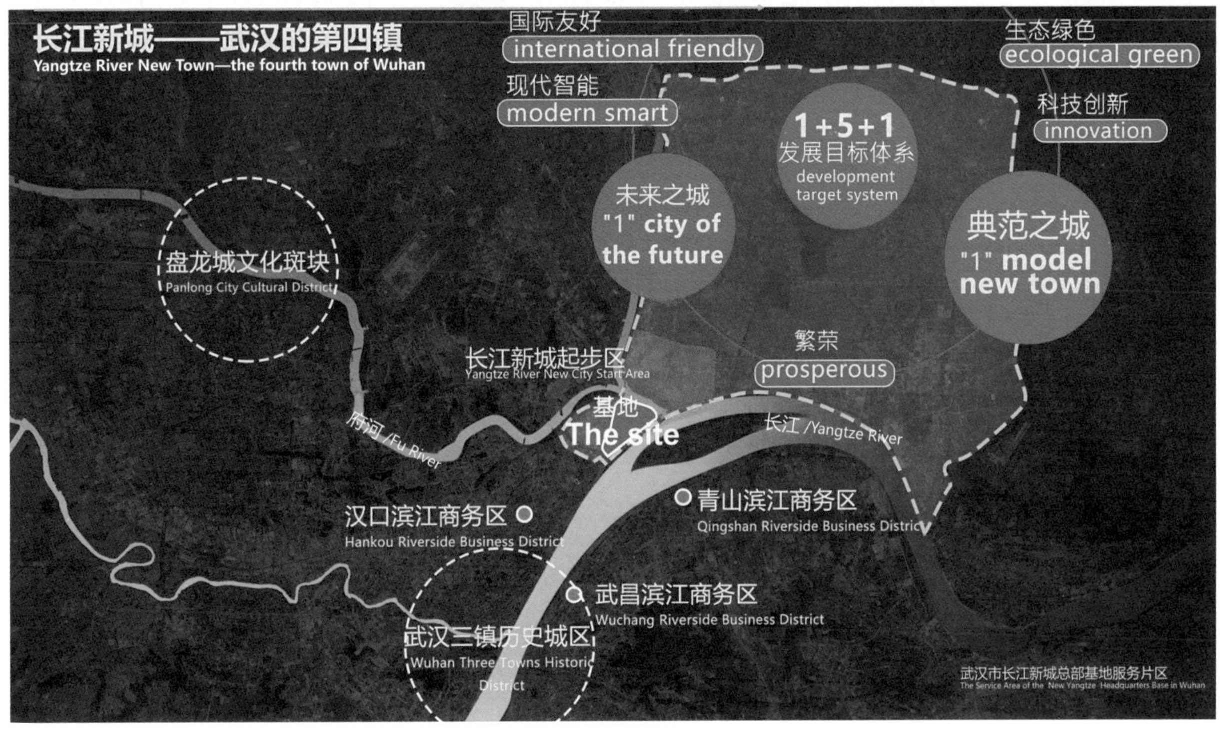

区位分析 Location Analysis

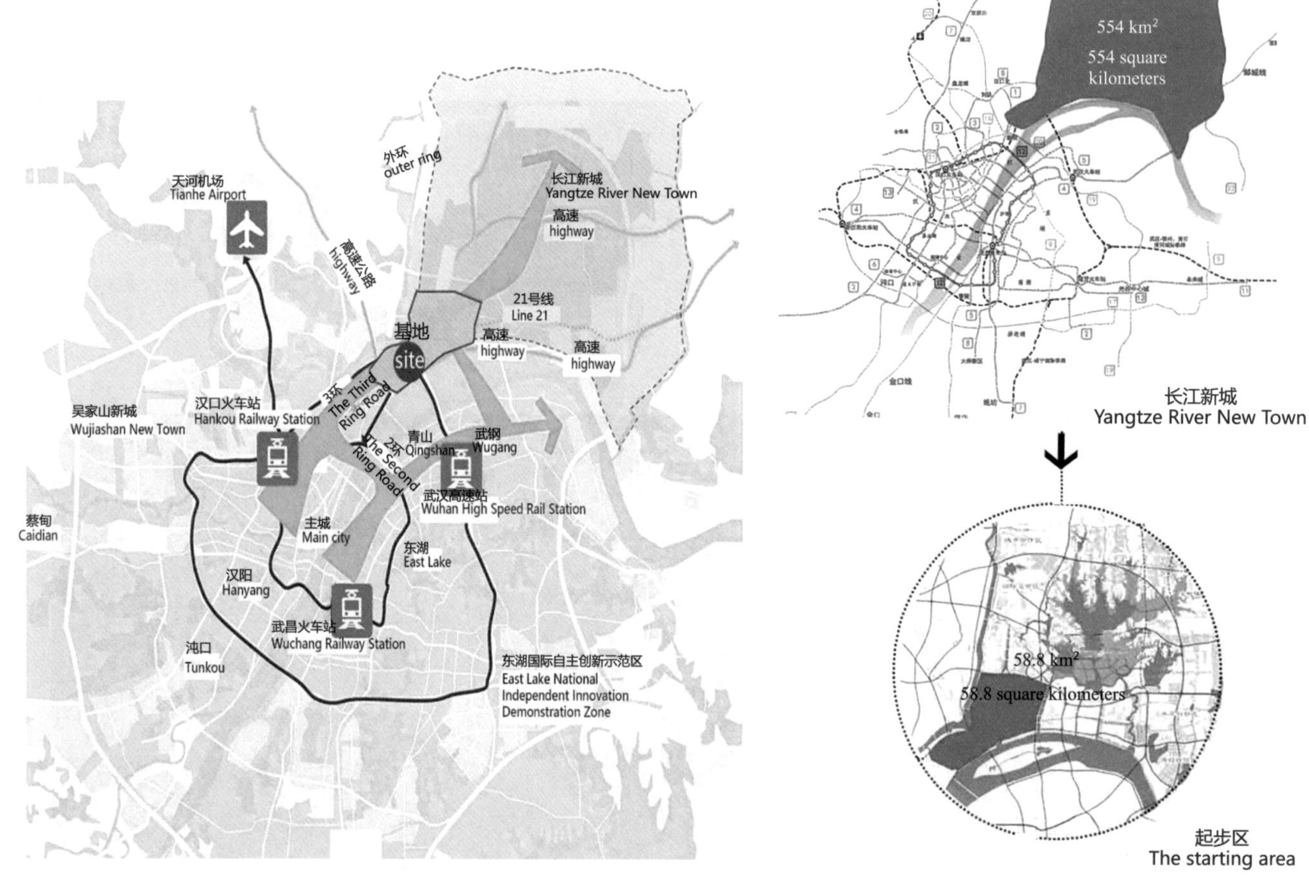

长江新城起步区位于中国湖北省武汉市，跨越江岸区、黄陂区 2 个行政区。规划用地东至武湖泵站河，南至长江北岸，西至滠水河、府河，西南至张公堤路，北至四环线。

The starting area of Yangtze River New Town is located in Wuhan, Hubei Province, China, spanning two administrative regions, Jiang'an District and Huangpi District. The planned land extends to the Wuhu Pumping Station River in the east, the north bank of the Yangtze River in the south, the Sheshui River and the Fu River in the west, Zhanggongdi Road in the southwest, and the Fourth Ring Road in the north.

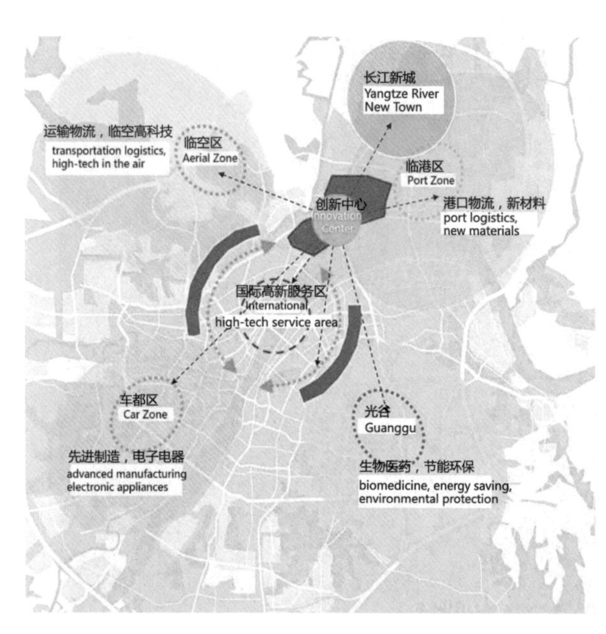

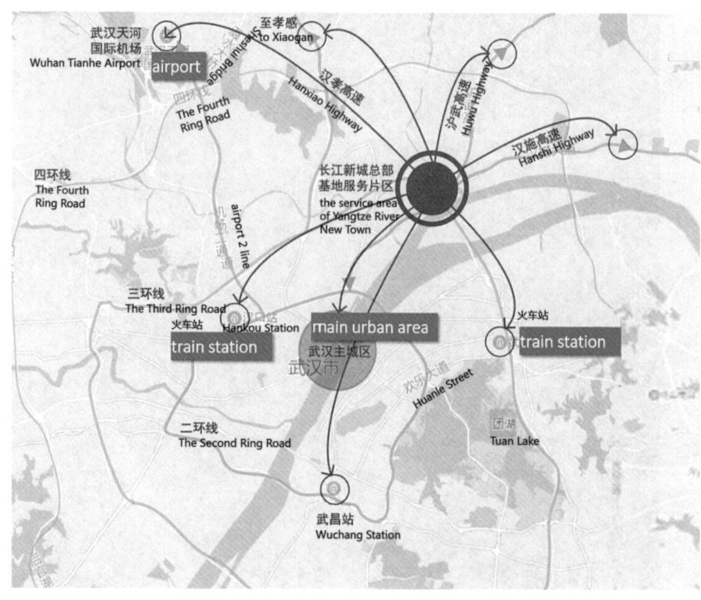

场地现状分析 Site Analysis

道路层级 Roads Hierarchy

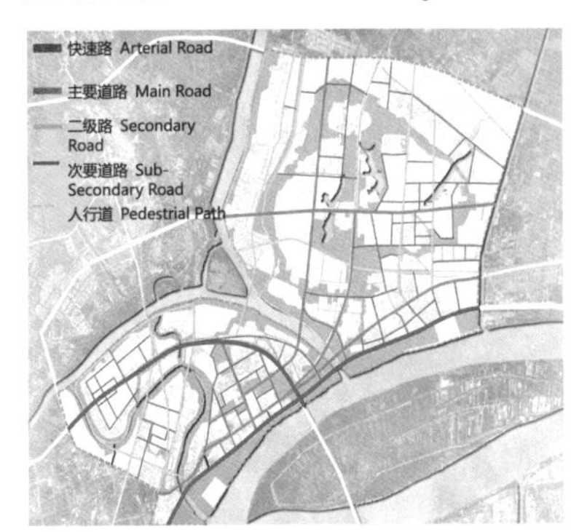

基地位于武汉市三环北端，区位交通条件优越，主要以朱家河、府河、长江及周边高速环绕，区块面积约900hm^2。

The Site is located in the northern end of the Third Ring Road of Wuhan, with superior location and traffic conditions, mainly surrounded by Zhujia River, Fu River, Yangtze River and surrounding highways. And the block area is about 900 hectares.

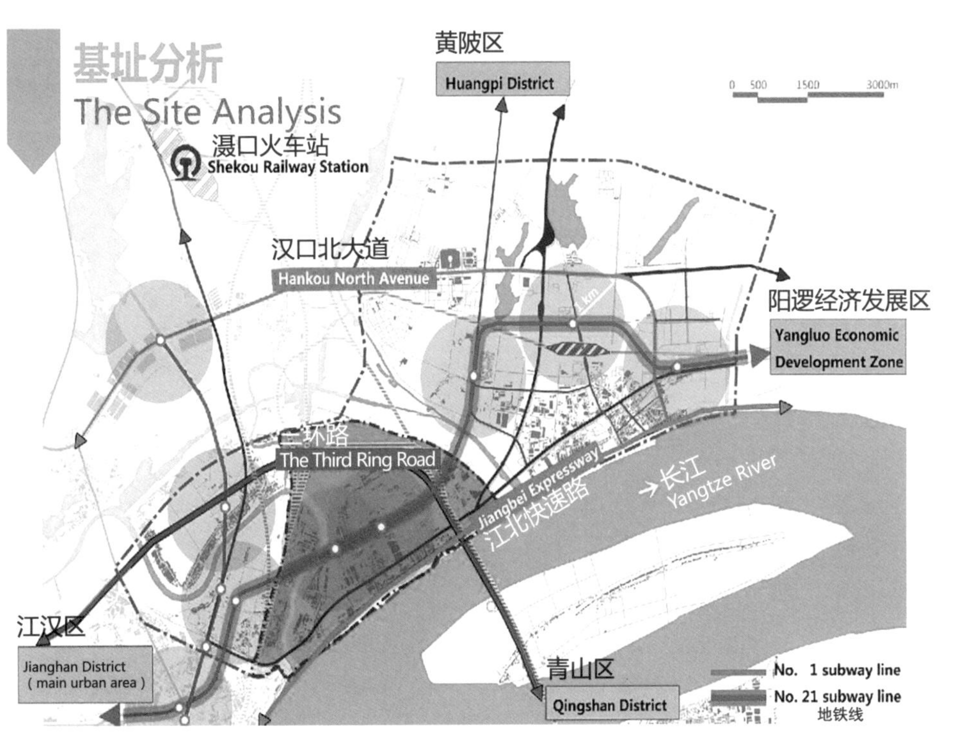

水体分析 Water analysis

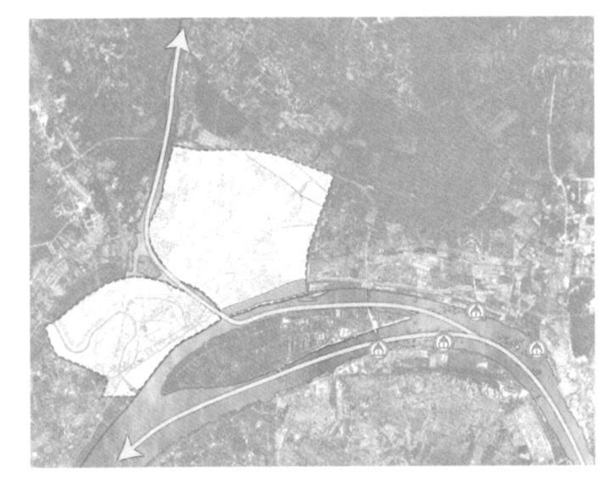

建筑情况 Building conditions

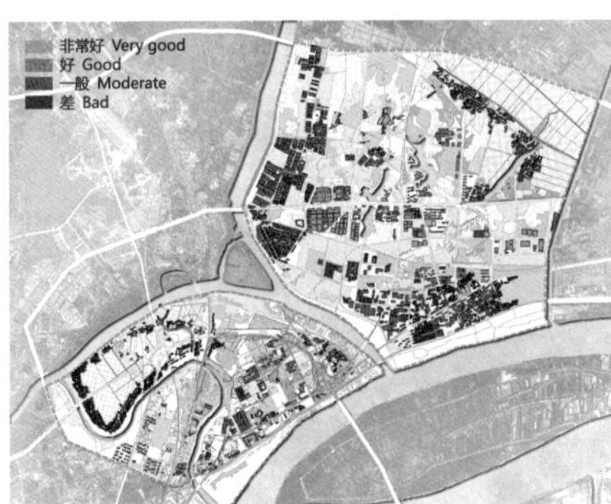

建筑肌理 Building texture

恒定水位 Constant water level

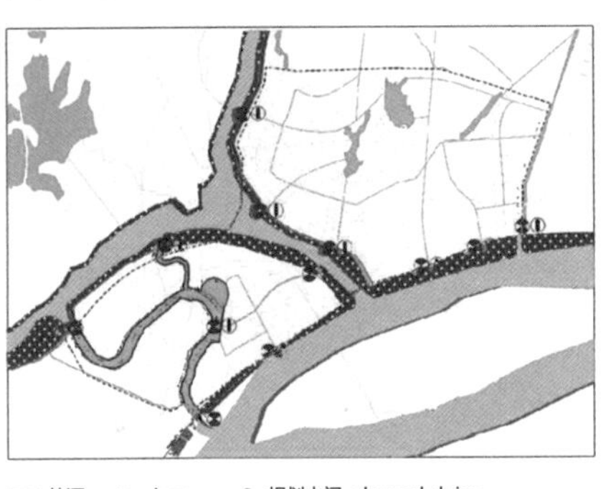

一般洪水位 General flood level

洪水位 Flood stage

文化分析 Cultural Analysis

红色文化 Red Culture

辛亥革命爆发后，国民军与清军在谌家矶多次交战，黄兴、成炳荣是主要的历史人物；战争时期军队多次驻扎于此；同时这里也曾组织抗日救亡运动。

After the outbreak of the Revolution of 1911, the National Army and the Qing Army fought many times in Chenjiaji, with Huang Xing and Cheng Bingrong as the main historical figures. The army stationed here many times during the Civil War between the Kuomintang and the Communist Party. At the same time, there were scenes of anti-Japanese and national salvation.

工业文化 Industrial culture

1861 年至清末民初，谌家矶一直为工业重地，1905 年建立芦汉铁路谌家矶车站；1907 年于此设立扬子机器制造有限公司，诞生了第一个中国造的川江炮舰；1909 年设立谌家矶造纸厂；1984 年，创立武汉铝厂。

From 1861 to the end of Qing Dynasty and the beginning of the Republic of China, Chenjiaji has always been an important industrial area. In 1905, Chenjiaji Stati on of Luhan Railway was established; In 1907, Yangzi Machinery Manufacturing Co., Ltd. was established here, giving birth to the first state-made Chuanjiang gunboat; In 1909, Chenjiaji Paper Mill was established; In 1984, Wuhan Aluminum Factory was established.

谌家矶
Chenjiaji

堤角工业区
（1958年武汉市工业区分布图）
Dijiao Industrial Zone
（Distribution map of Wuhan's industrial zones in 1958）

Five Dynasties and Ten Kingdoms 五代十国

梁武帝萧衍，曾于今武汉东北之谌家矶一带派兵一举击溃官军。

Emperor Wu of Liang, Xiao Yan, once sent troops in the area of Chenjiaji in the northeast of Wuhan at present to defeat the official army in one fell swoop.

Three Kingdoms 三国

赤壁之战时，刘备军队退军经过谌家矶。

During the Battle of Chibi, Liu Bei's army retired and passed through Chenjiaji.

Chu Culture 楚文化

在汉口东北郊谌家矶附近出土的我国铜剑、陶器等，属战国时期楚墓，是中华人民共和国成立后武汉市第一次发现楚墓与楚文物。

Warring States bronze sword pottery was unearthed near Chenjiaji in the northeastern suburb of Hankou. It belongs to the tomb of Chu during the Warring States period. It is the first time that Chu tombs and cultural relics have been discovered in Wuhan after the founding of the People's Republic of China.

Name source 名称来源

早年有谌姓人家在此捕鱼和生活，之后便慢慢出现了以姓氏命名的矶，称为谌家矶。

In the early years, people with the surname Chen fished and lived here, and then gradually appeared the Ji named after the surname, called Chenjiaji.

Industry 工业

1861年到清末民初，武汉基本形成沿长江汉口谌家矶造纸、机械工业区，租界食品加工带，刘家庙交通运输设备工业区。

From 1861 to the end of the Qing Dynasty and the beginning of the Republic of China, Wuhan basically formed a paper-making and machinery industrial zone along the Yangtze River in Hankou Chenjiaji, a concession food processing zone, and a Liujiamiao transportation equipment industrial zone.

红色文化 Red culture

1911年辛亥革命爆发，10-11月期间黄兴、成炳荣与清军多次在谌家矶交战，登陆谌家矶，进攻刘家庙是军队的主要作战策略。

1920年7月6日，冯玉祥部统率全军，移驻谌家矶造纸厂。

1920年7月，吉鸿昌随十六混成旅移防武汉谌家矶。

1927年元月，贺龙率国民革命军队开抵武汉，一部驻扎在谌家矶。

In 1911, the Revolution of 1911 broke out. From October to November, Huang Xing, Cheng Bingrong and the Qing Army fought and landed in Chenjiaji many times. Attacking Liujiamiao was the main choice of the army.

On July 6, 1920, Feng Yuxiang took over the whole army and moved to Chenjiaji Paper Mill.

In July 1920, Ji Hongchang moved to Wuhan Chenjiaji with the 16th Mixed Brigade.

In January 1927, He Long led a team of the National Revolutionary Army to Wuhan, and the army was stationed in Chenjiaji.

红色文化 Red culture

1938年“青救”基层组织成立，并至谌家矶一带宣传抗日救亡形势。

1938年蒋介石在谌家矶设立据点，9月，第五十七师撤离至汉口谌家矶。

In 1938, the "Qing Salvation" grassroots organization was established, and it spread the form of anti-Japanese national salvation in Chenjiaji area.

In 1938, Chiang Kai-shek set up a stronghold in Chenjiaji. In September, the 57th Division hold fast and evacuated to Chenjiaji.

码头文化 Wharf culture

凭借汉口商业发展，谌家矶沿岸一直船满为患，直至1965年武湖围垦、府河改道、下游河床淤塞，民船货运才转向长江。中华人民共和国成立前后，境内有6条不定期的航线，包括谌家矶往返汉口、堤角的航线。

With the commercial development of Hankou, the coast of Chenjiaji was always in a state of being overwhelmed with ships. Until 1965, the Wu Lake was reclaimed, the Fu River was rerouted, the downstream riverbed was blocked, and freight transport by civilian ships turned to the Yangtze River. Before and after liberation, there were 6 irregular routes in the territory, including the routes from Chenjiaji to Hankou and Dijiao.

1655 — 1905 — 1907 — 1931 — 1984

N S W E

图例 legend

高：0.996513 high

低：0 low

0 0.75 1.5 3 km

植被覆盖度 Vegetation Coverage

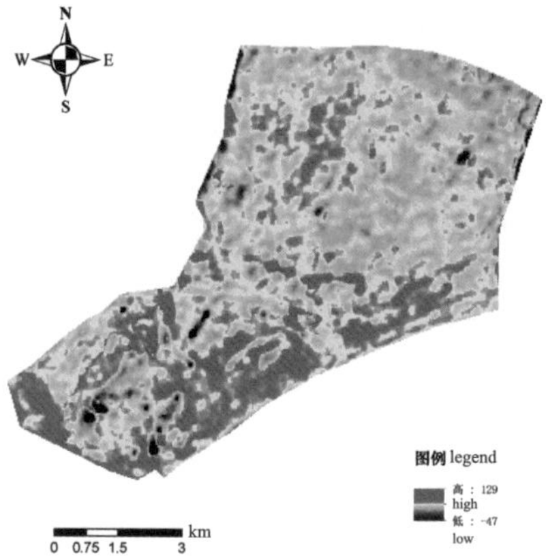

N S W E

图例 legend

高：129 high

低：-47 low

0 0.75 1.5 3 km

高程 Elevation

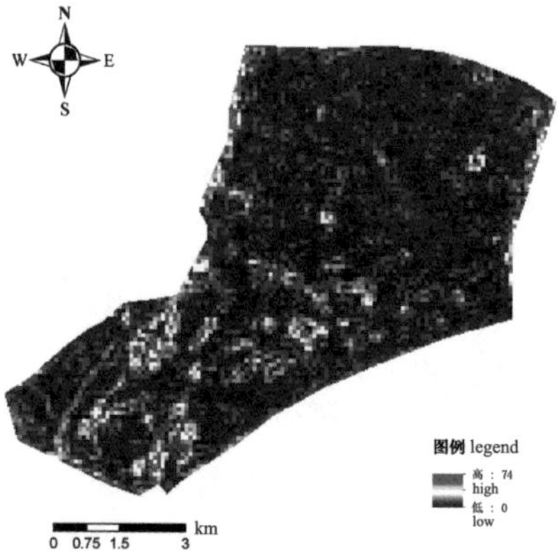

N S W E

图例 legend

高：74 high

低：0 low

0 0.75 1.5 3 km

地形起伏度 Terrain Undulate

第一部分　中国学生作品

Works from China

蓝绿渗透
The Infiltration of Blue-Green

卢亚雯
LU Yawen

李蓝
LI Lan

夏文莹
XIA Wenying

赵芊芊
ZHAO Qianqian

苑竟达
YUAN Jingda

谢德灵
XIE Deling

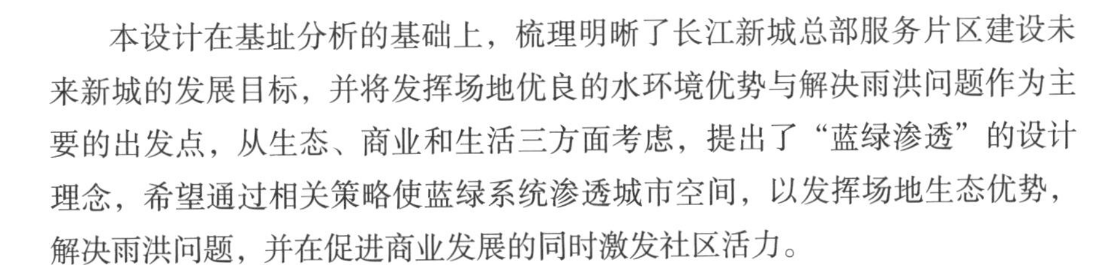

本设计在基址分析的基础上，梳理明晰了长江新城总部服务片区建设未来新城的发展目标，并将发挥场地优良的水环境优势与解决雨洪问题作为主要的出发点，从生态、商业和生活三方面考虑，提出了“蓝绿渗透”的设计理念，希望通过相关策略使蓝绿系统渗透城市空间，以发挥场地生态优势，解决雨洪问题，并在促进商业发展的同时激发社区活力。

Based on the analysis of the base site, this design clarified the development goals of the service area of the Yangtze River New City Headquarters to build a new city in the future, and took the advantage of the excellent water environment of the site and the solution of rain and flood problems as the main starting point. Considering the three aspects of ecology, business and life, the design concept of “The Infiltration of Blue-Green” was proposed, hoping to make the blue-green system penetrate the urban space through related strategies, so as to give full play to the ecological advantages of the site, solve the rain-flood problem, and stimulate the community while promoting commercial development vitality.

概念设计 Concept Design

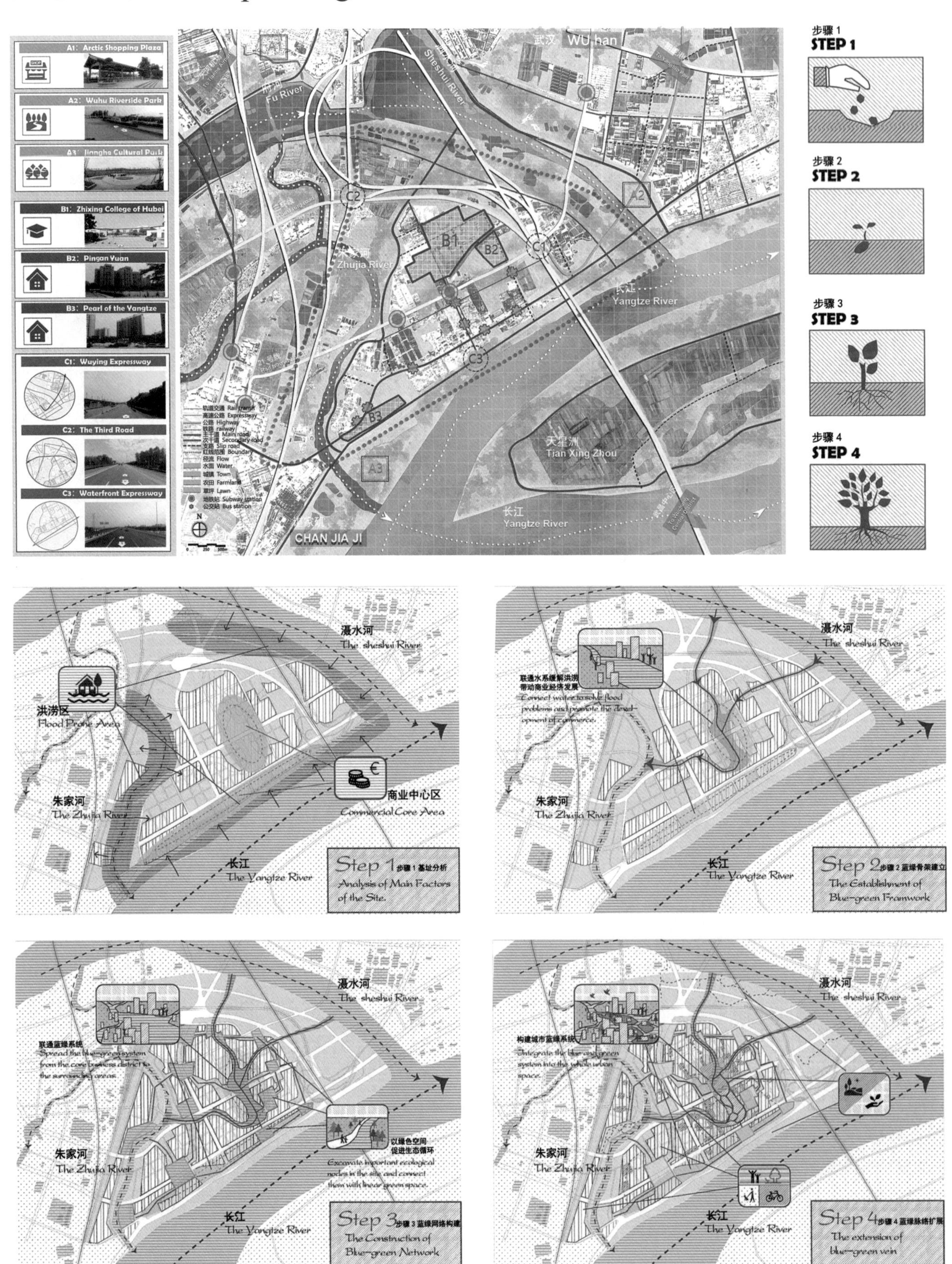

设计策略 Design Strategy

设计策略分为三个层次，包括生态策略、商业策略和生活策略。在生态方面削弱热岛效应，强化雨洪管理。在商业方面构建共享街道和重视低影响开发。在生活方面从宏观到微观，使蓝绿系统渗入城市—社区—绿地。

The design strategy is divided into three levels, including ecological strategy, business strategy and life strategy. In the ecological aspect, weaken the heat island effect and strengthen rain-flood management. On the commercial side, build shared streets and emphasize low-impact development. From the macro to the micro in life, the blue-green system infiltrates the city-community-green space.

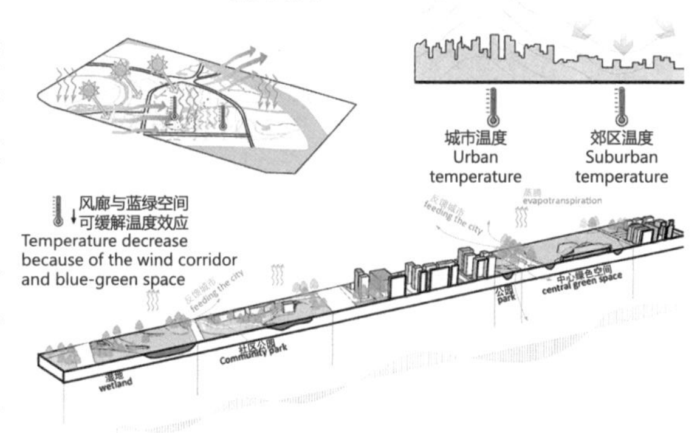

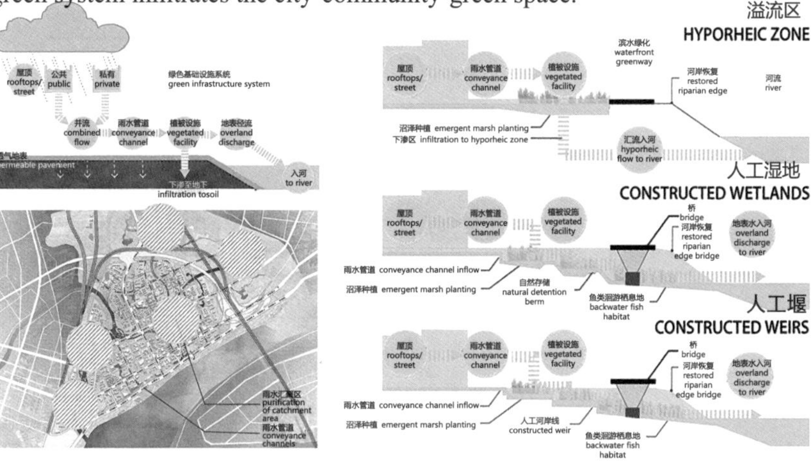

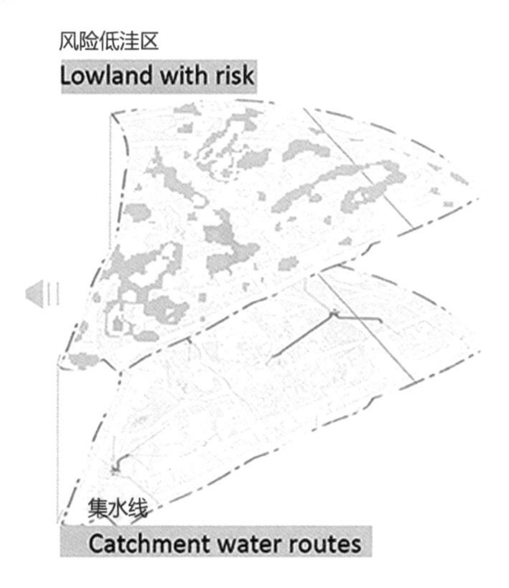

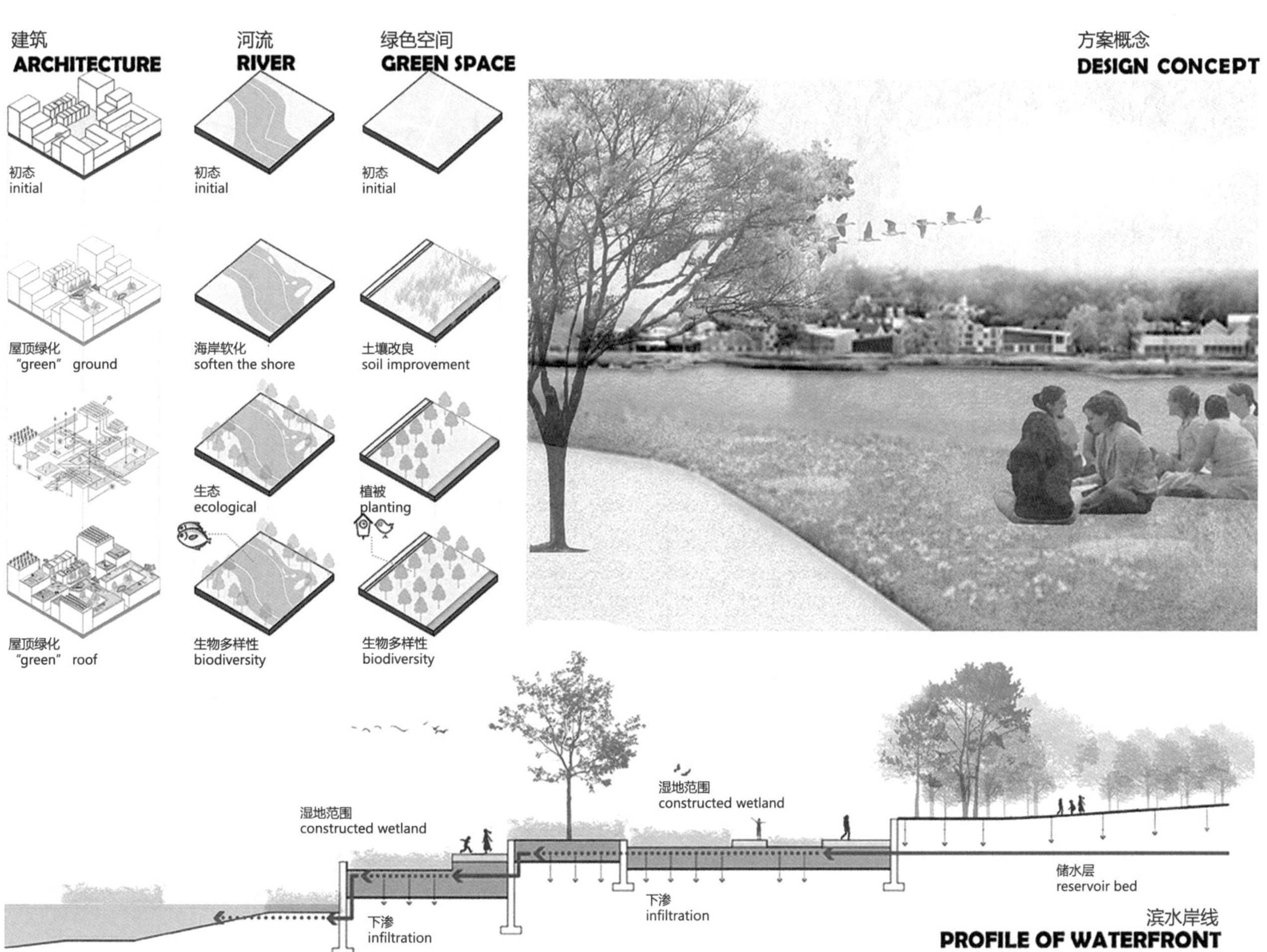

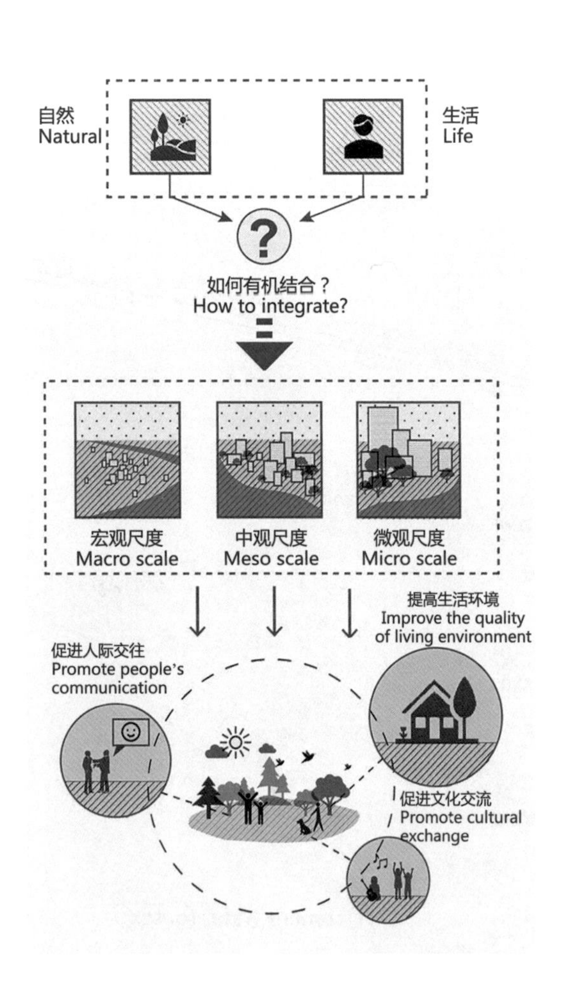

自然
Natural
生活
Life
如何有机结合？
How to integrate?
宏观尺度
Macro scale
中观尺度
Meso scale
微观尺度
Micro scale
提高生活环境
Improve the quality of living environment
促进人际交往
Promote people's communication
促进文化交流
Promote cultural exchange

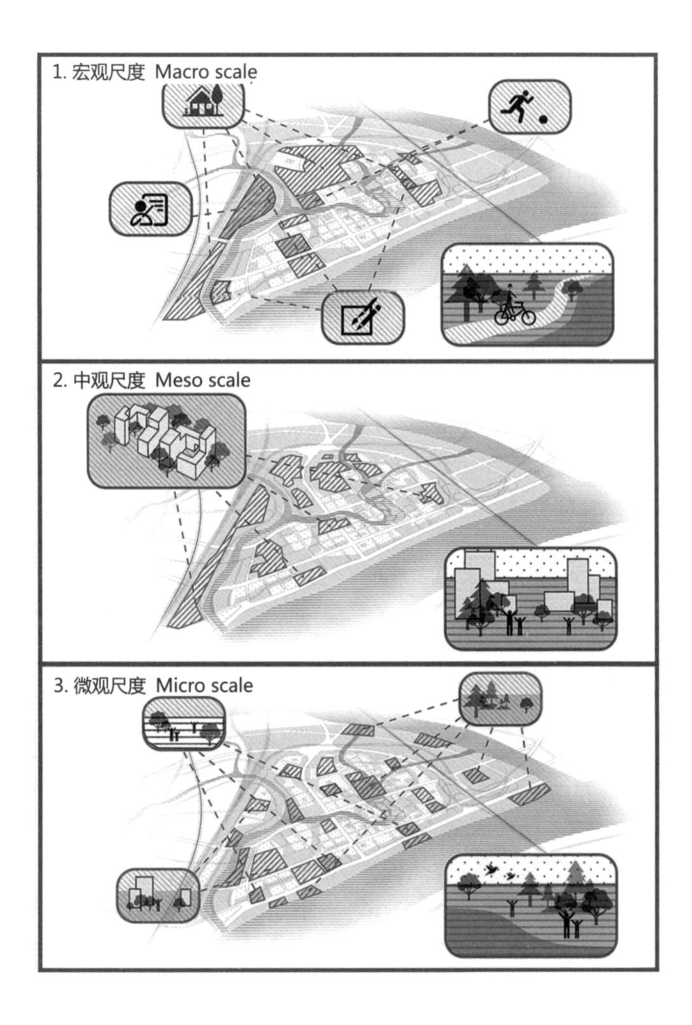

1. 宏观尺度 Macro scale
2. 中观尺度 Meso scale
3. 微观尺度 Micro scale

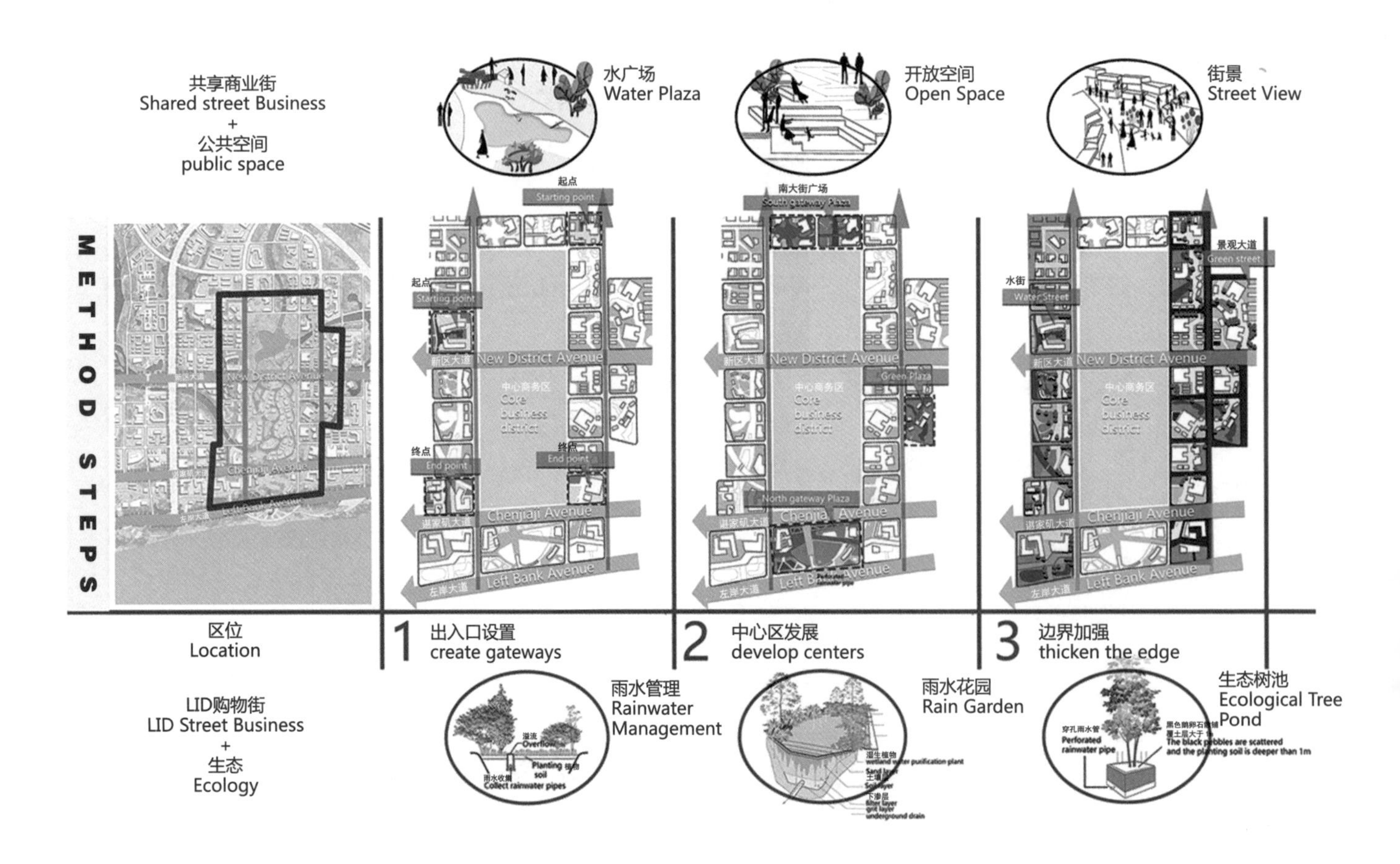

共享商业街
Shared street Business
+
公共空间
public space
水广场
Water Plaza
开放空间
Open Space
街景
Street View
METHOD STEPS
New District Avenue
Chenjiaji Avenue
起点
Starting point
终点
End point
新区大道 New District Avenue
中心商务区
Core business district
陈家矶大道 Chenjiaji Avenue
左岸大道 Left Bank Avenue
南大街广场
South gateway Plaza
Green Plaza
North gateway Plaza
景观大道
Green street
水街
Water Street
区位
Location
1 出入口设置
create gateways
2 中心区发展
develop centers
3 边界加强
thicken the edge
LID购物街
LID Street Business
+
生态
Ecology
雨水管理
Rainwater Management
雨水花园
Rain Garden
生态树池
Ecological Tree Pond
溢流
Overflow
Planting soil
雨水收集
Collect rainwater pipes
穿孔雨水管
Perforated rainwater pipe

设计分析 Design Analysis

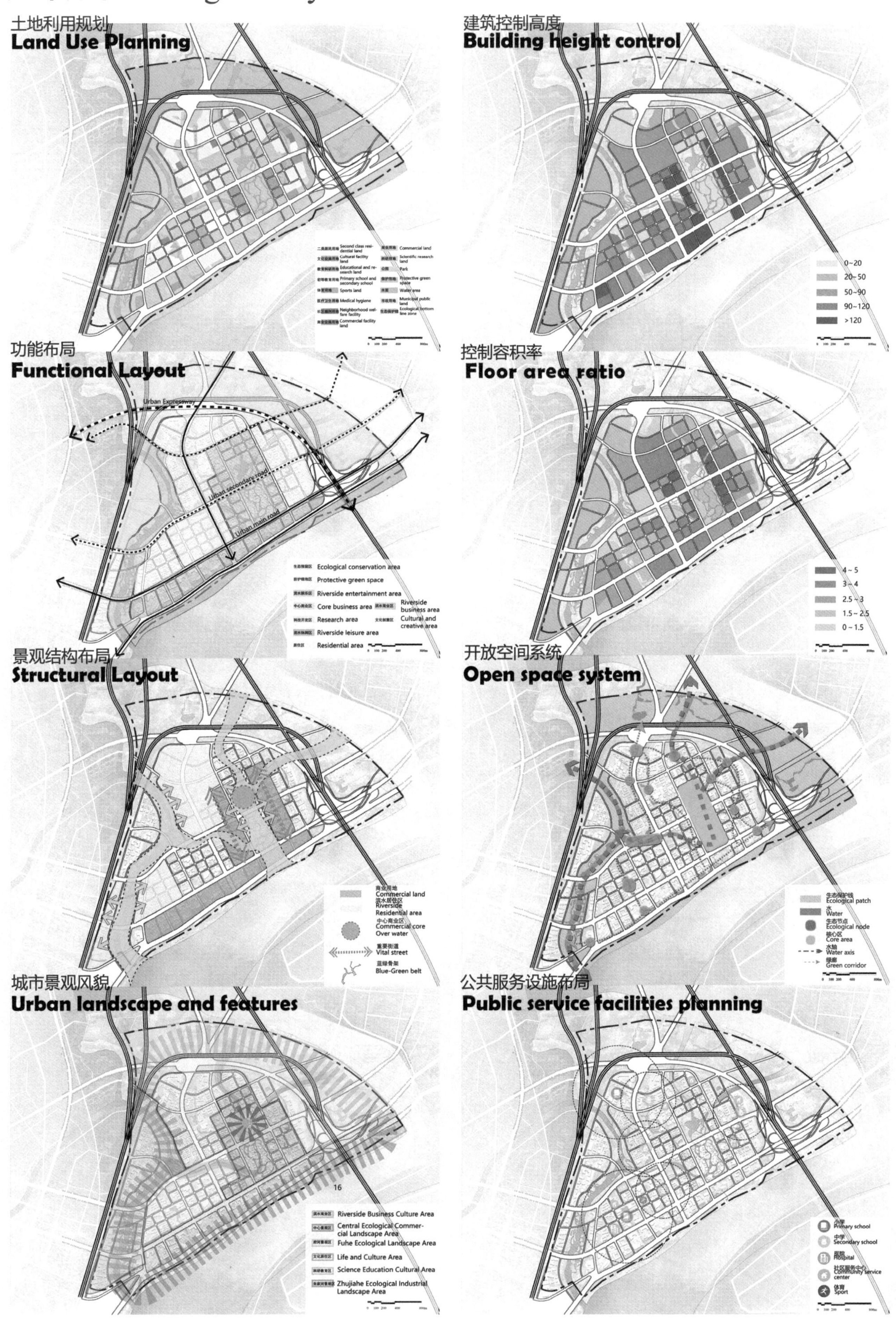

天河机场
To Tianhe Airpo
乡野风光带
Farmland Scenery
滨水平台
Waterfront Platform
滨水漫步道
Slow Waterfront
商业圈
Business Circle
谌家矶中学
Chenjiaji Middle School
朱家河
Zhujia River
景观大道
Green Street
滨水购物公园
Waterfront Shopping Park
生态社区
Introverted ecological community
地下隧道
Underground Tunnel
科技中心
Research Center
购物岛
Vibrant Shopping Islands
体育场
Stadium
水街
Water Street
商业广场
Commercial Plaza
文创区
Cultural and Creative District
教育基地
Educational Base
朱家河湿地
Zhujiahe Wetland
朱家河绿地
Zhujiahe Greenland
生活圈
Life Circle
汉口中心区
To Hankou Central District
长江
Yangtze River
国际研发中心
WORLD DESIGN STUDIO

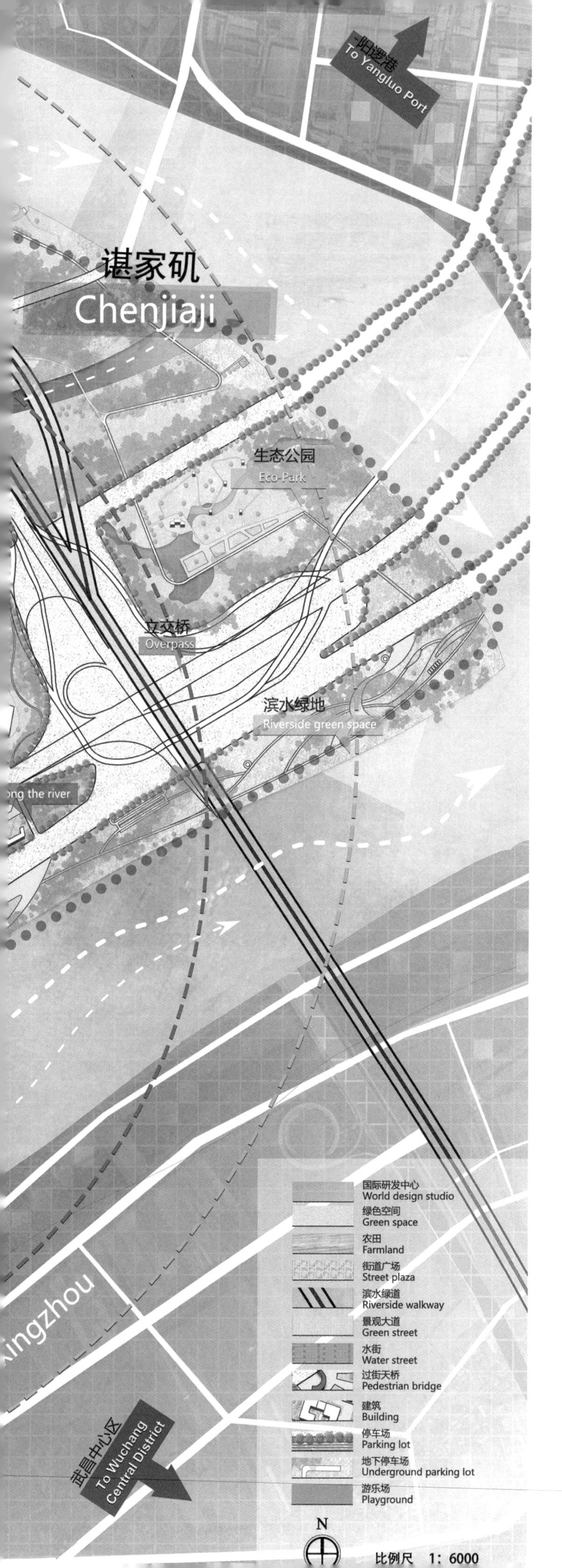

结合基址分析总结出的场地的机遇与挑战，本次设计以蓝绿渗透为设计理念进行城市设计。从生态、商业、生活三个层面出发，通过雨洪管理、滨水商业商务区、慢行系统、内向型社区等，将蓝绿系统渗透进入城市空间以改善生态环境，解决洪水问题，激发商业活力，提升宜居住性。

Combined with the site's opportunities and challenges summarized by the base site analysis, this design takes the blue and green penetration as the design concept for urban design. Starting from the three levels of ecology, commerce and life, through strategies such as stormwater management, waterfront commercial and business districts, slow-moving systems, and introverted communities, the blue-green system is penetrated the urban space to improve the ecological environment, solve the flood problem, and stimulate commercial vitality to enhance livability.

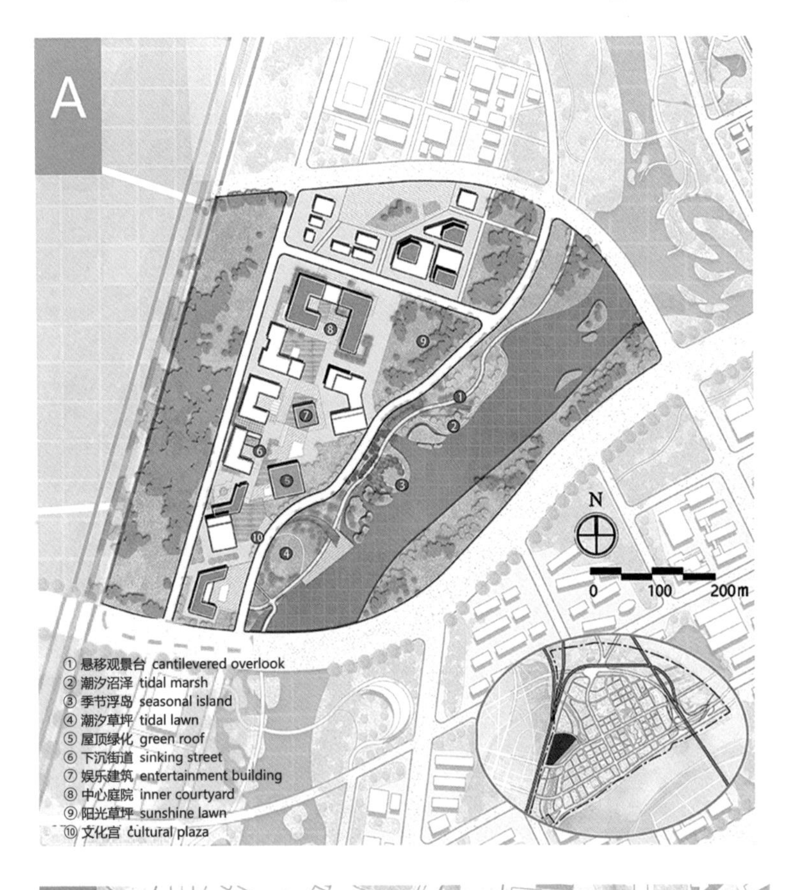

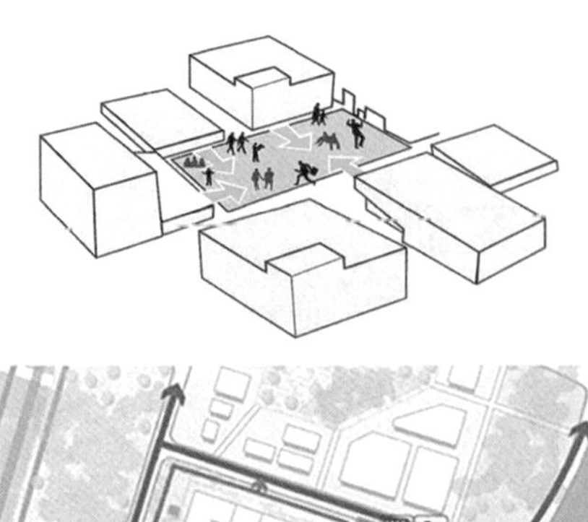

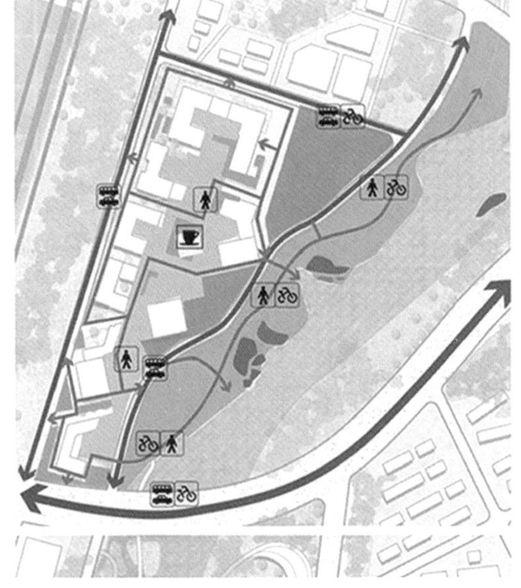

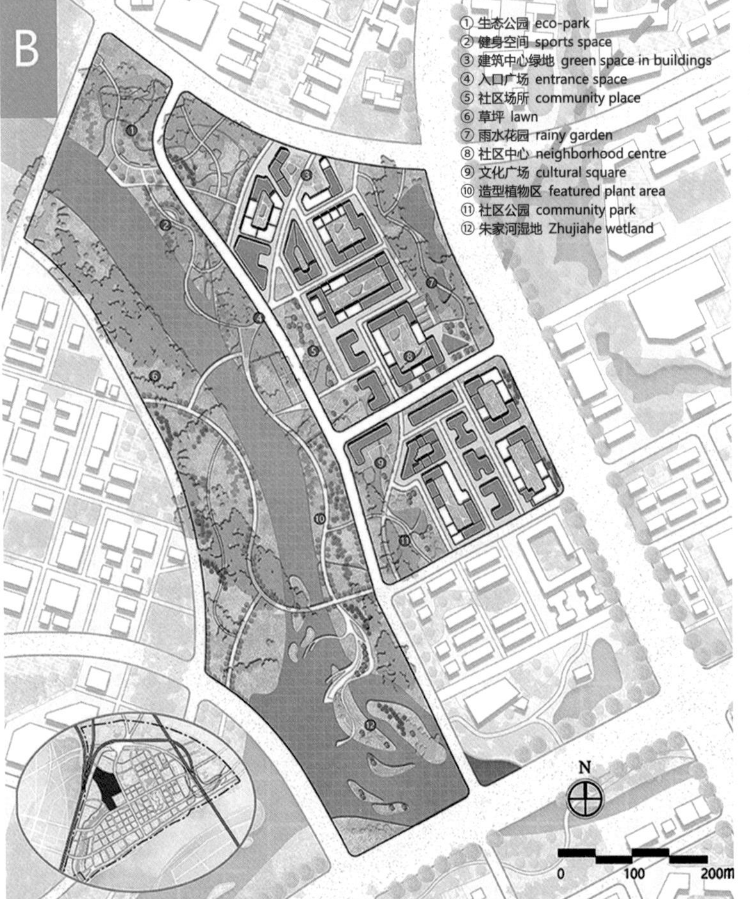

区域 A: 科技产业园区。业余交流和“思维狂潮”是产品和服务上的突破性创新产生的重要过程。因此，我们的目标是在园区内营造激发思考创新和促进参与合作的灵活公共空间。

区域 B: 内向型社区。引水至居住区内以强化水的渗透。水体形态丰富，沿河景观设计提供公共空间从而激发人与水的联系，增加该区域的自然性和娱乐性。

Area A: Science and technology industrial park. Amateur communication and “thinking frenzy” are important processes for the production of breakthrough innovations in products and services. Therefore, our goal is to create a flexible public space in the park that stimulates thinking and innovation and promotes participation and cooperation.

Area B: Introverted community. Diverting water to the residential area to strengthen the water penetration. The water is rich in forms. The riverside landscape provides public space to stimulate the connection between people and water and increase the naturalness and entertainment of the area.

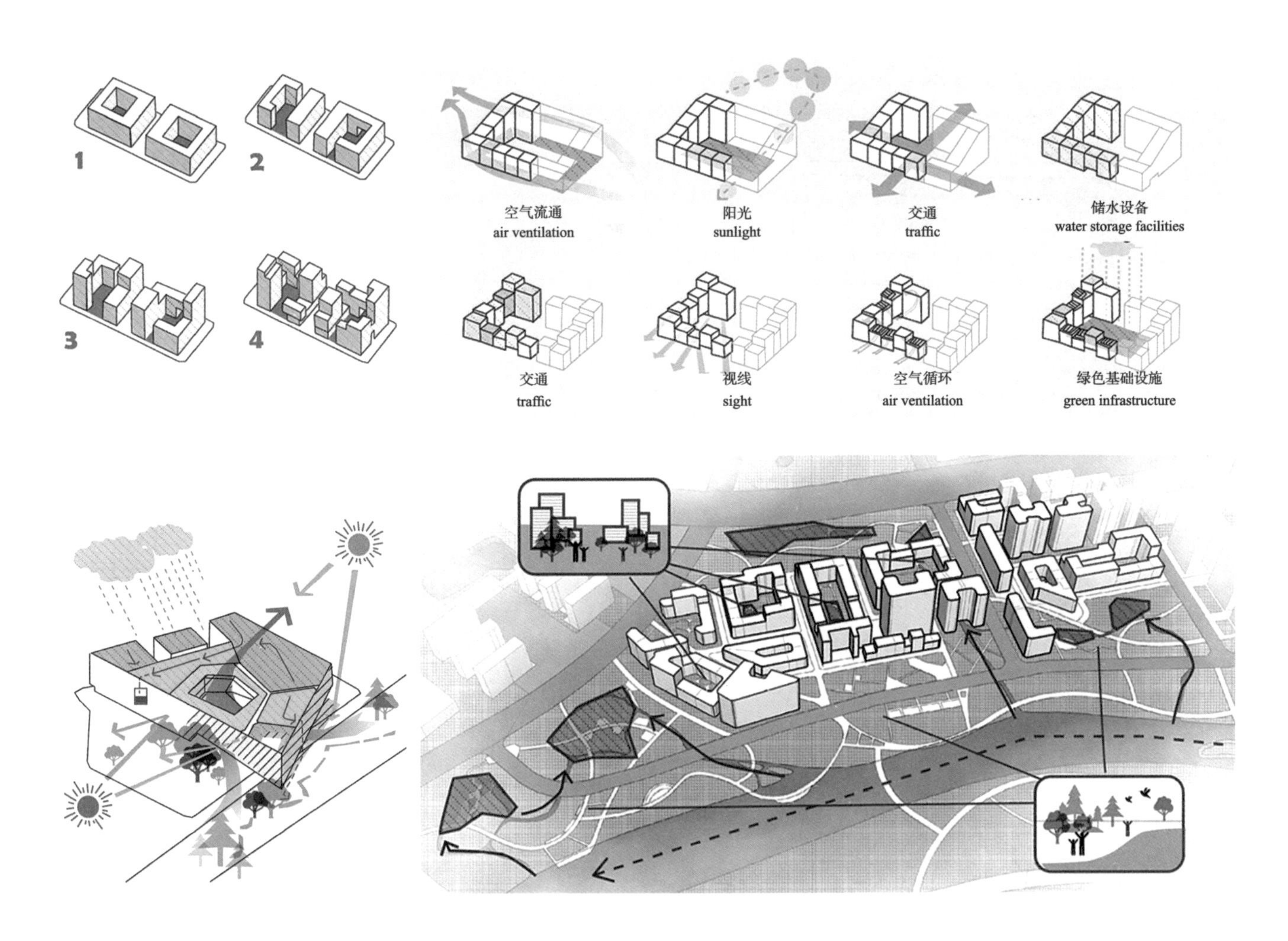
1
2
3
4
空气流通
air ventilation
阳光
sunlight
交通
traffic
储水设备
water storage facilities
交通
traffic
视线
sight
空气循环
air ventilation
绿色基础设施
green infrastructure

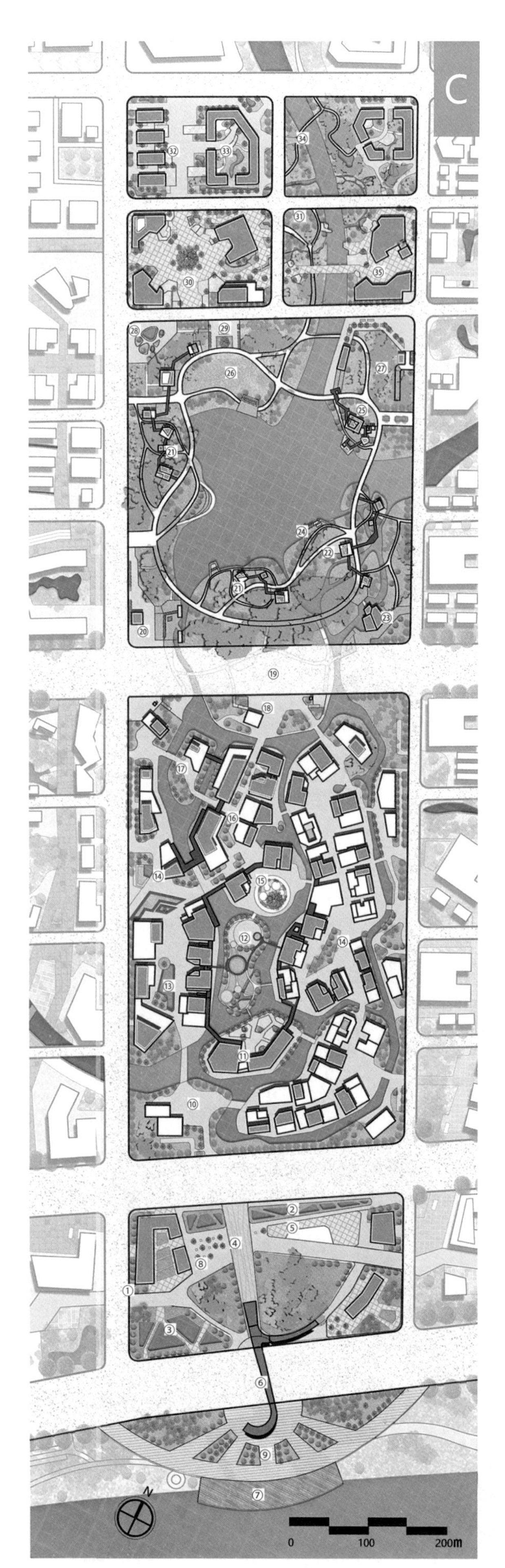

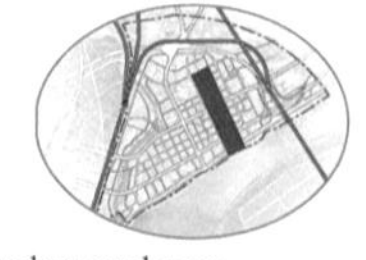

① 办公建筑 business office building
② 季节性泳池 seasonal pool
③ 镜面池 mirror pool
④ 特殊步道 special paving
⑤ 室外咖啡 outdoor cafe
⑥ 过街天桥 pedestrian bridge in the sky
⑦ 观景台 viewing platform
⑧ 树阵广场 tree array square
⑨ 入口广场 entrance plaza
⑩ 地下通道入口 underpass entrance
⑪ 空中商业走廊 aerial commercial street
⑫ 零售岛 retail island
⑬ 喷泉广场 fountain square
⑭ 雕塑广场 sculpture plaza
⑮ 圆形剧场 amphitheatre
⑯ 文化商业街 cultural commercial street
⑰ 主题酒店 theme hotel
⑱ 入口广场 entrance plaza
⑲ 地下空间 underground space
⑳ 地铁站 subway station
㉑ 空中商业 air commercial group
㉒ 生态湿地 ecological wetland
㉓ 购物大厦 shopping hall
㉔ 滨水平台 waterfront platform
㉕ 水塔 watchtower
㉖ 草坪 lawn
㉗ 桥 bridge
㉘ 街旁绿地 green space by the street
㉙ 公园入口 park entrance
㉚ 商业广场 commercial plaza
㉛ 滨水绿地 riverfront green space
㉜ 建筑群 unit building
㉝ 内环居住区 introverted community
㉞ 绿地广场 green plaza
㉟ 商业文化广场 business culture plaza

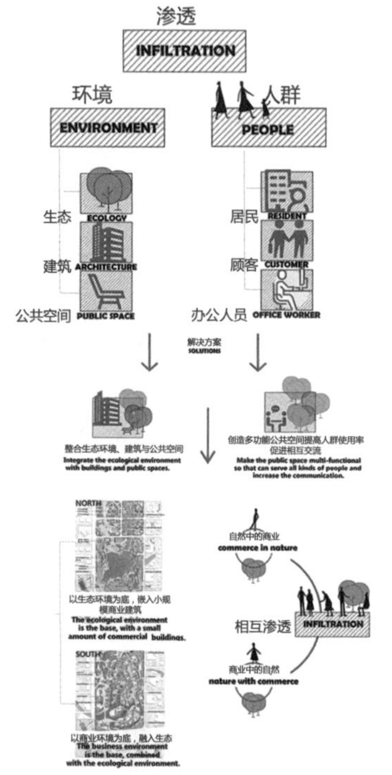

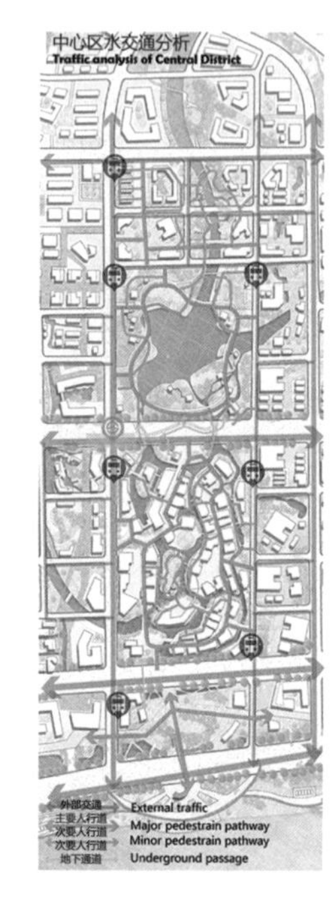

中心商业区位于城市中轴线上，是城市商业、生态、生活的核心区域，且独具特色。

区域内的滨水购物公园兼具生态休闲与商业功能，活力商业群岛内商业步行街与城市河流景观相结合，商业休闲广场以城市阳台连接滨江公园。

具有相对独立的交通系统，通过密集的慢行交通网络与外部交通连接。

The central business district is located on the central axis of the city and is the core area of the city's commerce, ecology and life, with unique characteristics.

The waterfront shopping park in the area has both ecological leisure and commercial functions. The commercial pedestrian street in the vibrant commercial archipelago is combined with the urban river landscape. The commercial leisure plaza is connected to the riverside park with an urban balcony.

Have a relatively independent transportation system. Connect with external traffic through a dense slow traffic network.

详细设计 Detailed Design

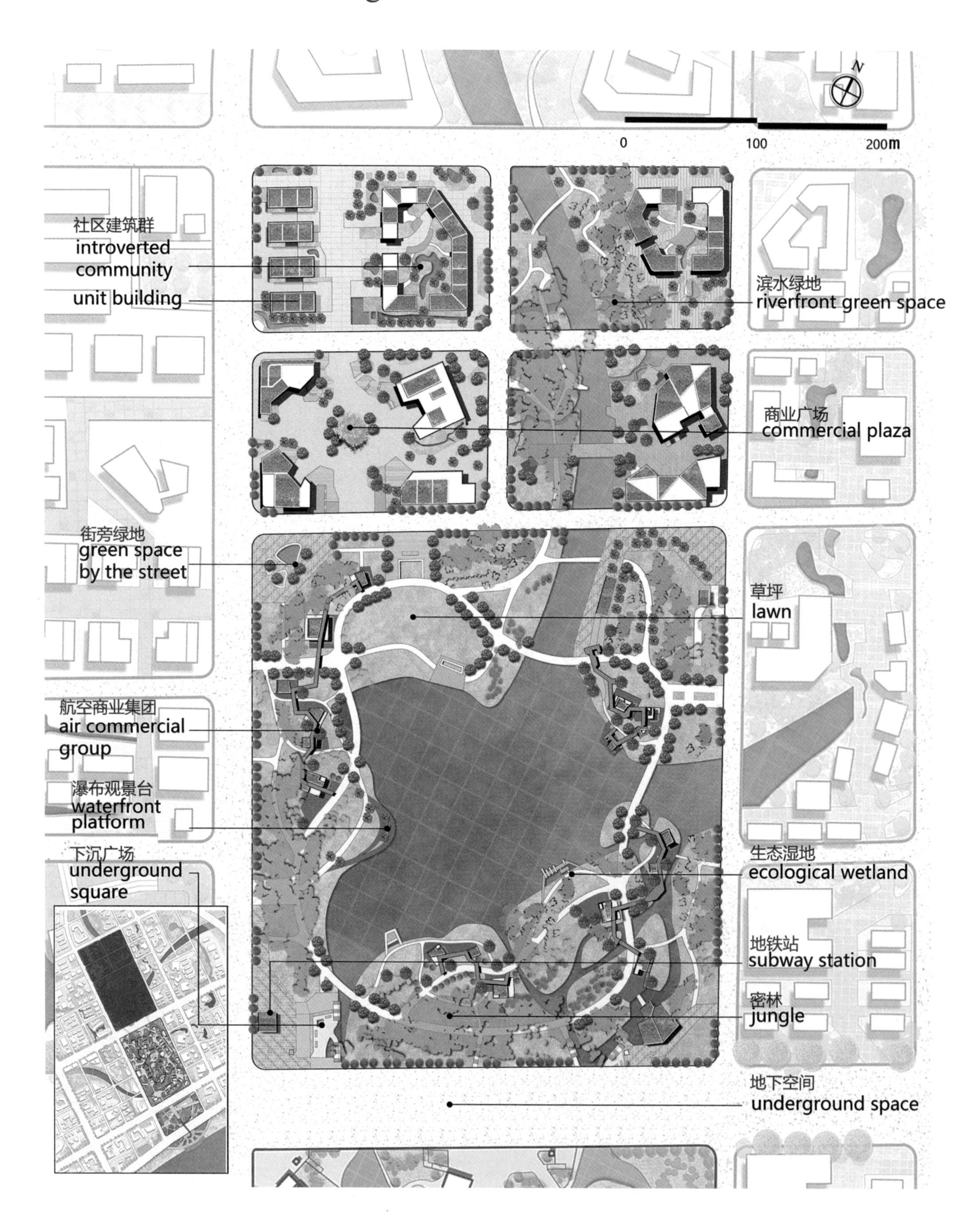

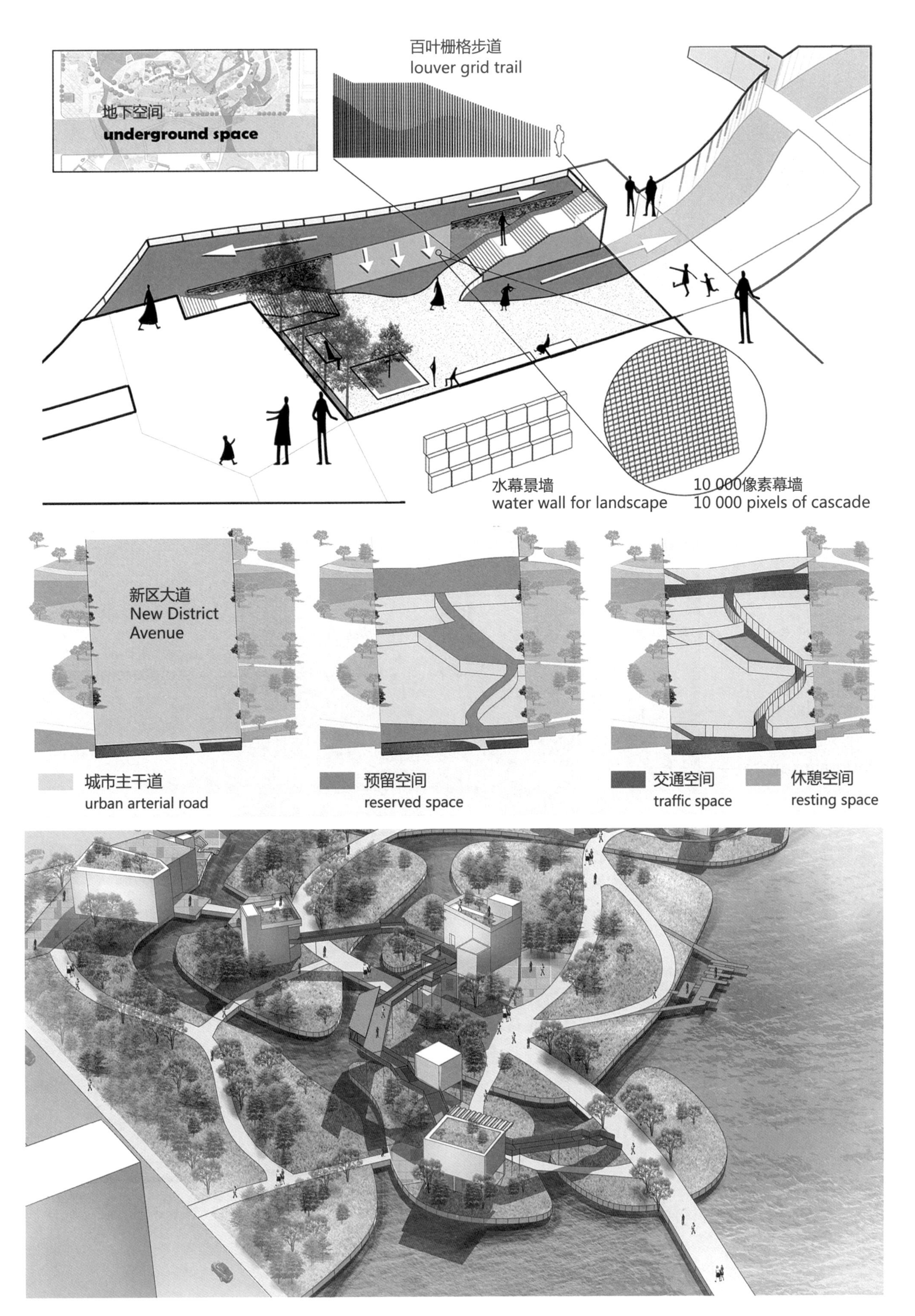

地下空间
underground space
百叶栅格步道
louver grid trail
水幕景墙
water wall for landscape
10 000像素幕墙
10 000 pixels of cascade
新区大道
New District
Avenue
城市主干道
urban arterial road
预留空间
reserved space
交通空间
traffic space
休憩空间
resting space

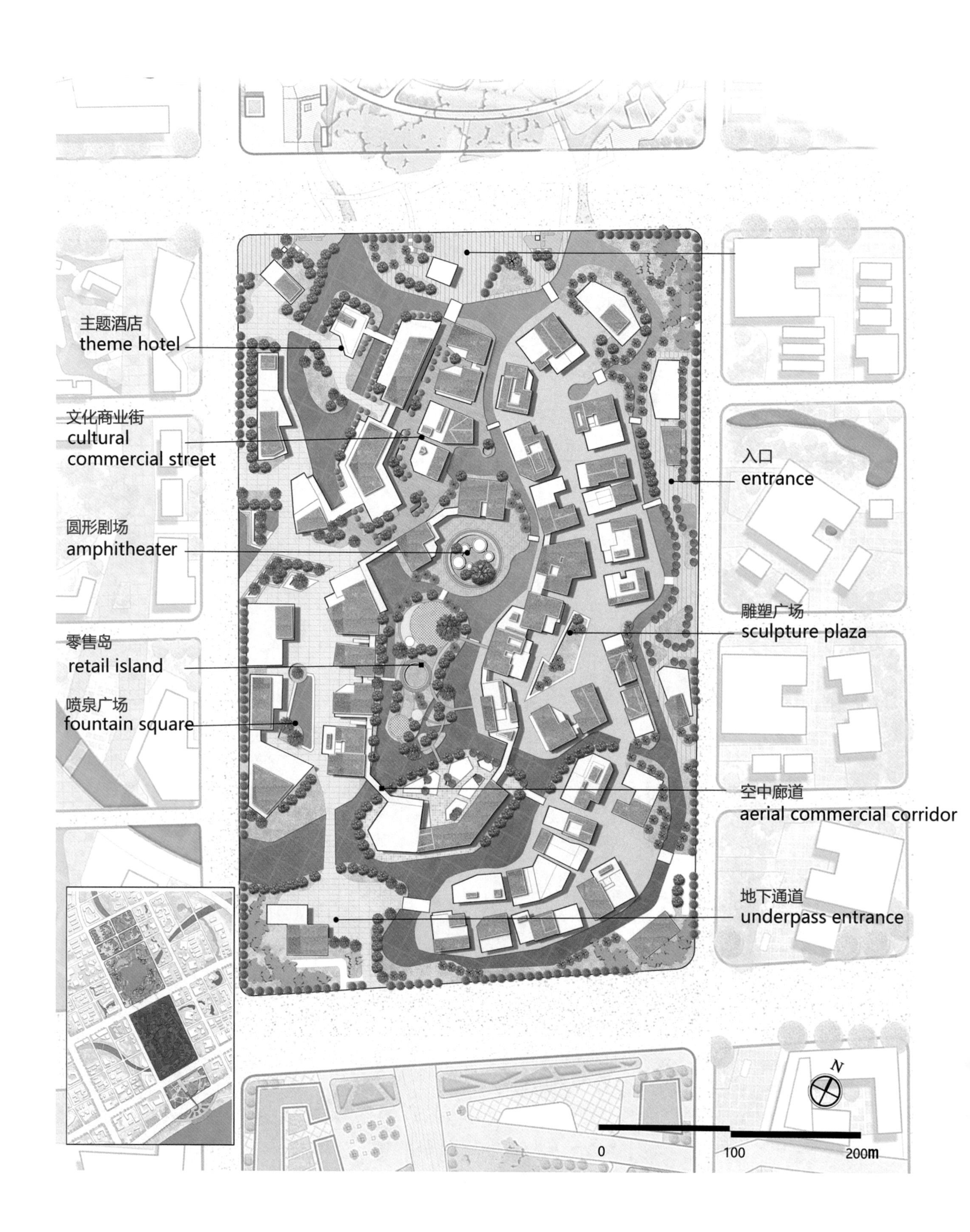
主题酒店
theme hotel
文化商业街
cultural
commercial street
圆形剧场
amphitheater
零售岛
retail island
喷泉广场
fountain square
入口
entrance
雕塑广场
sculpture plaza
空中廊道
aerial commercial corridor
地下通道
underpass entrance
N
0
100
200m

购物娱乐
pleasant shopping environment

顾客
customers

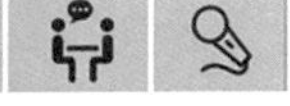

休闲娱乐行为
venues for occasional leisure activities

居民
residents

业余休闲娱乐
after-work recreation and rest

办公人员
officer workers

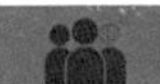

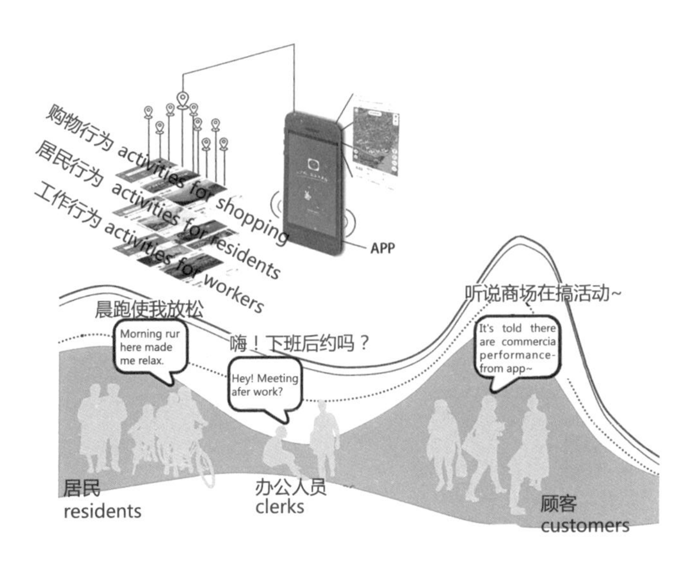

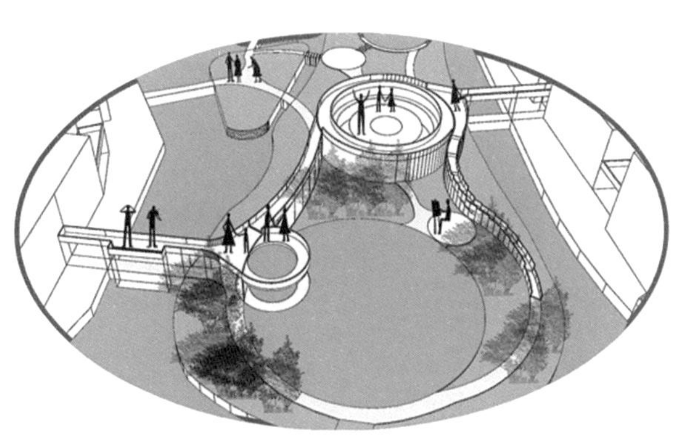

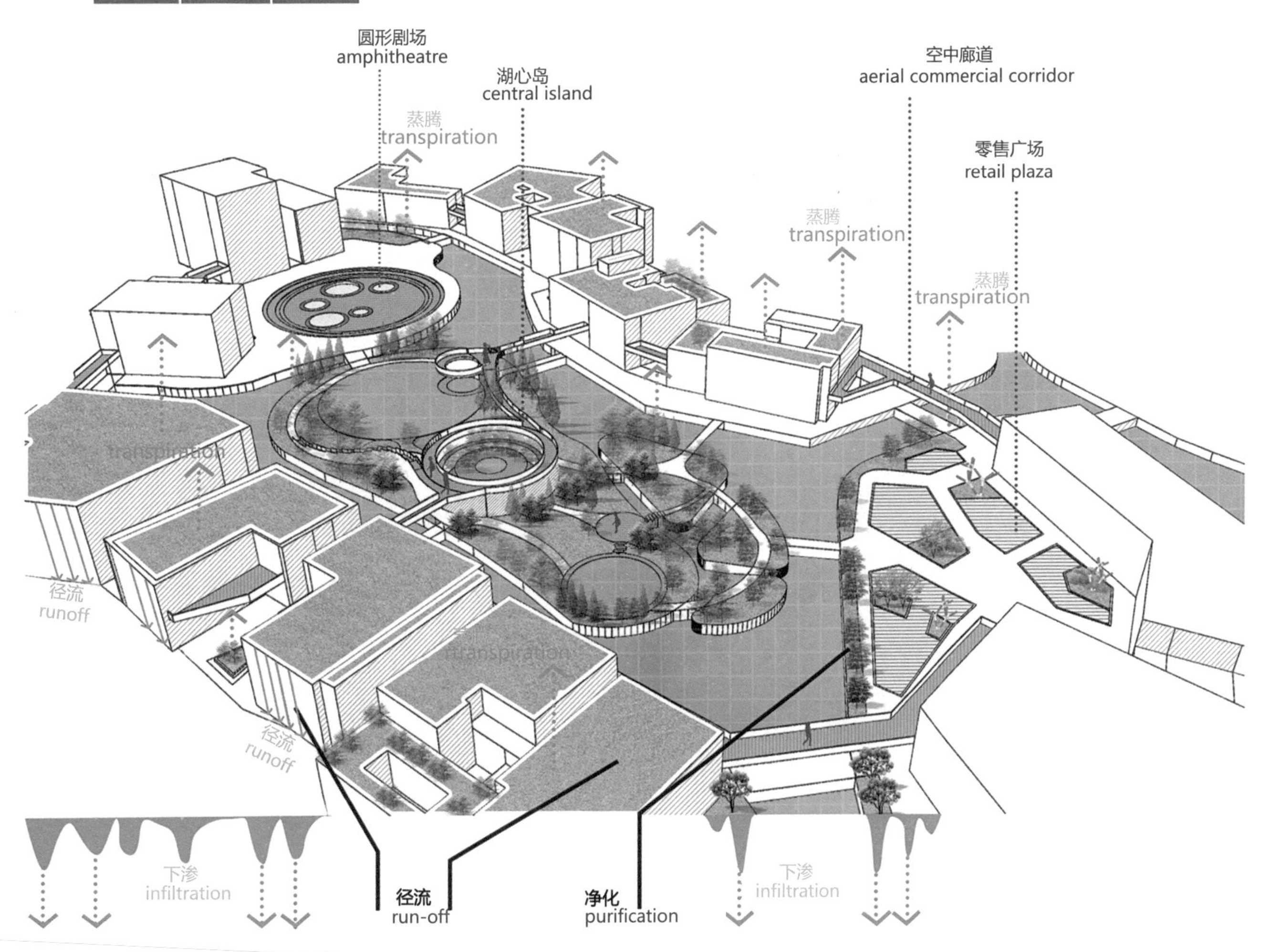

建筑策略

Strategy For Architecture

建筑结合景观空间&空中走廊
Architecture with green space & aerial corridor

三种建筑结合景观空间模式
Three models for architecture with green space

空间策略

Strategy For Public Space

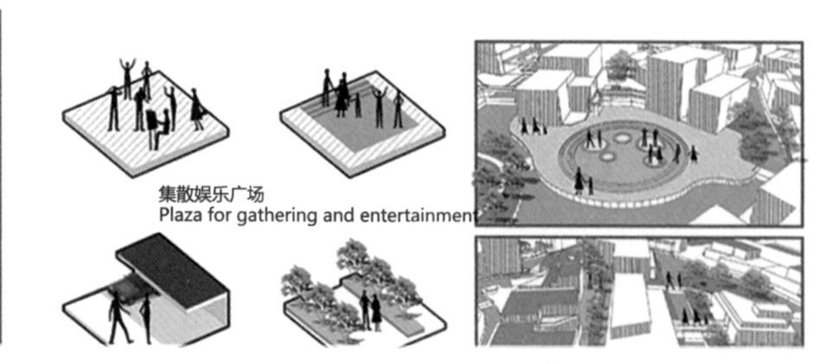

剖面图

Section

商业建筑 commercial building	街道 street	绿地 green space	河流（硬质驳岸） river (hard bank)	绿地 green space	商业建筑 commercial building

居民活动
Activities for residents

办公活动
Activities for workers

顾客活动
Activities for customers

综合活动
Activities for everyone

循水造境，蓝绿交融
—— 基于雨洪管理的弹性城市设计
Co-exist with nature
— Resilient urban design based on stormwater management

贺敏文
HE Minwen

许骄阳
XU Jiaoyang

黄子秋
HUANG Ziqiu

赵聆言
ZHAO Lingyan

关艺蕾
GUAN Yilei

谢婧
XIE Jing

周梓璇
ZHOU Zixuan

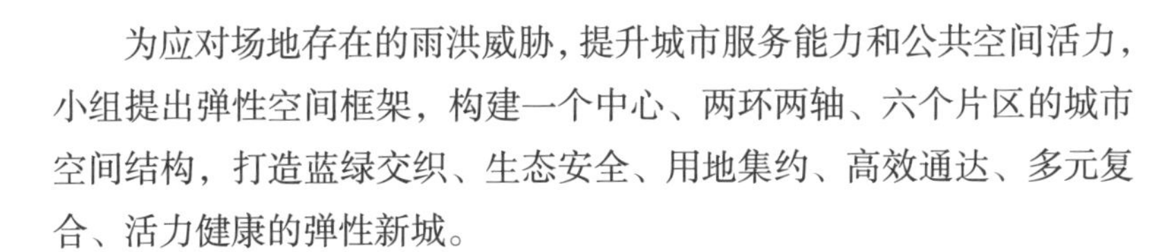
为应对场地存在的雨洪威胁，提升城市服务能力和公共空间活力，小组提出弹性空间框架，构建一个中心、两环两轴、六个片区的城市空间结构，打造蓝绿交织、生态安全、用地集约、高效通达、多元复合、活力健康的弹性新城。

In order to deal with the rain and flood threat existing in the site, improve the urban service ability and the vitality of public space, our group proposed an elastic spatial framework to build an urban spatial structure of one center, two rings, two axes, and six districts, so as to create a resilient new city with blue and green interwoven, ecological security, intensive land use, efficient and accessible, multiple and complex, vitality and healthy.

设计策略 Design Strategy

策略一 蓝绿交融

Strategy 1 Blends of blue and green

总体蓝绿安全格局 Overall blue-green security pattern

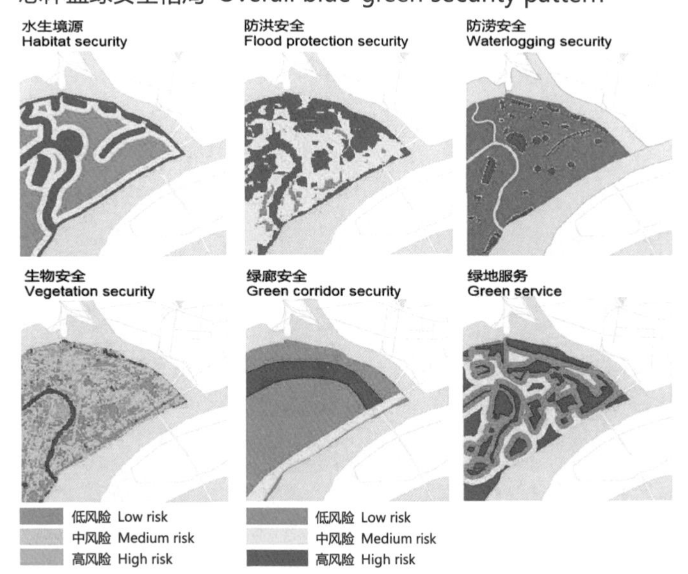

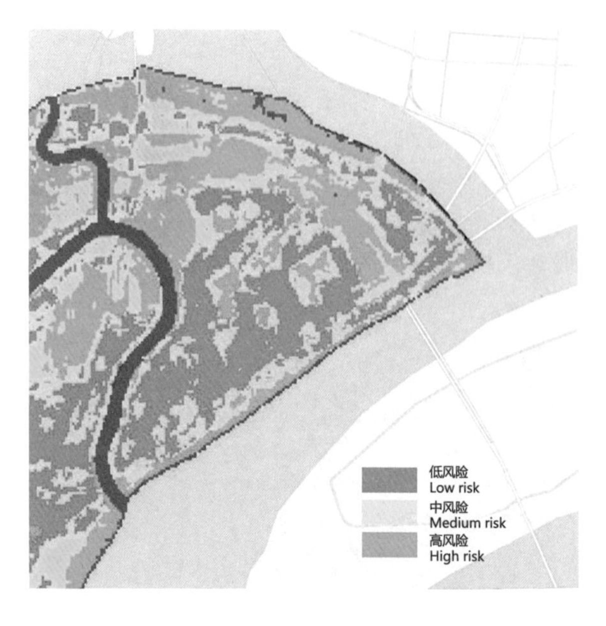

总体蓝绿空间规划 Overall blue and green space planning

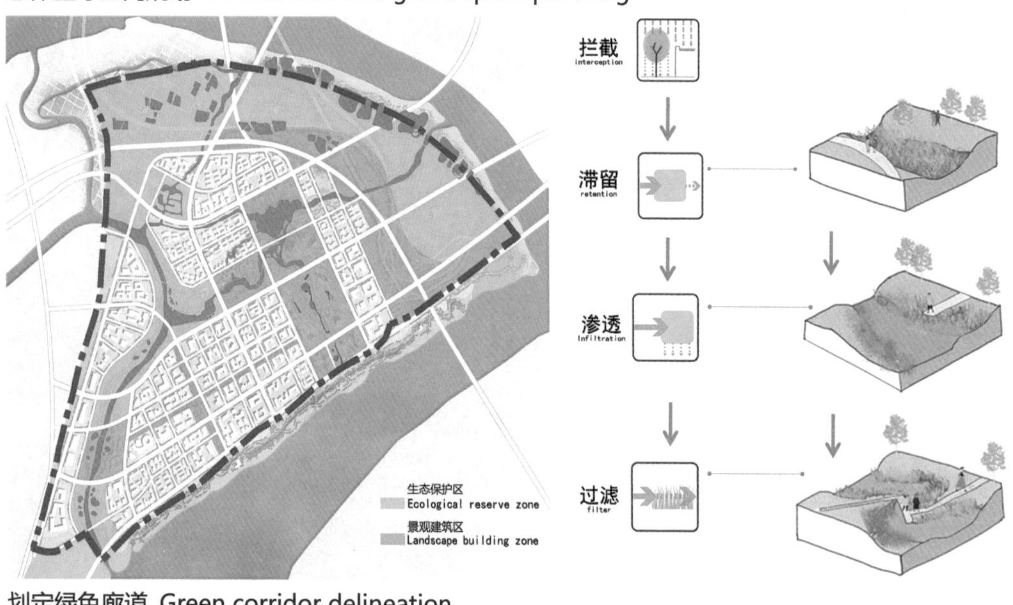

注重保护河道两侧原生态自然景观，利用自然河道，形成道路网络，构筑城镇骨架，将水系和绿地网络相结合，形成蓝绿空间系统。

Pay attention to the protection of the original ecological natural landscape on both sides of the river, use the natural river to form a road network, build the urban skeleton, and combine the water system and the green space network to form a blue-green space system.

划定绿色廊道 Green corridor delineation

一级走廊 500~700m
First-level corridor 500~700m

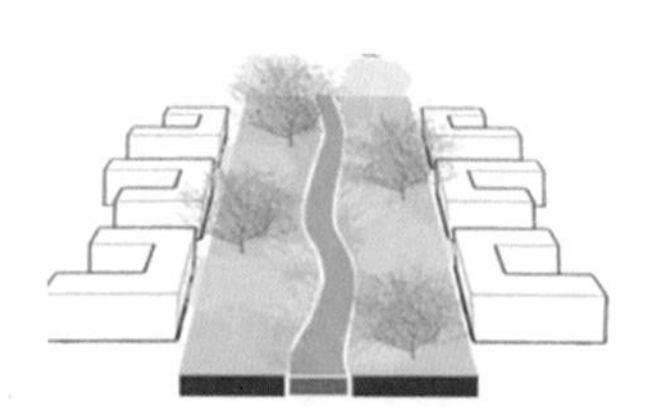

二级走廊 300~500m
Second corridor 300~500m

三级走廊 200~300m
Third-level corridor 200~300m

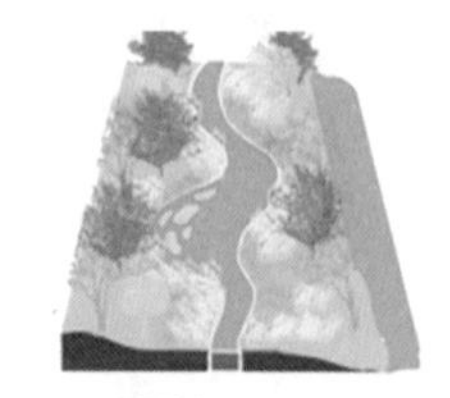

四级走廊 100~200m
Fourth-level corridor 100~200m

多形态蓝绿汇水设施 Multi-form blue and green catchment facilities

绿色空间布局 Green space layout　　蓝色空间布局 Blue space layout

汇水区内的水依据高差汇入片区内多种形态的蓝绿设施内。

The water in the catchment area flows into various forms of blue-green facilities according to the height difference.

灰色设施布局——水闸 Gray facility layout—Sluice

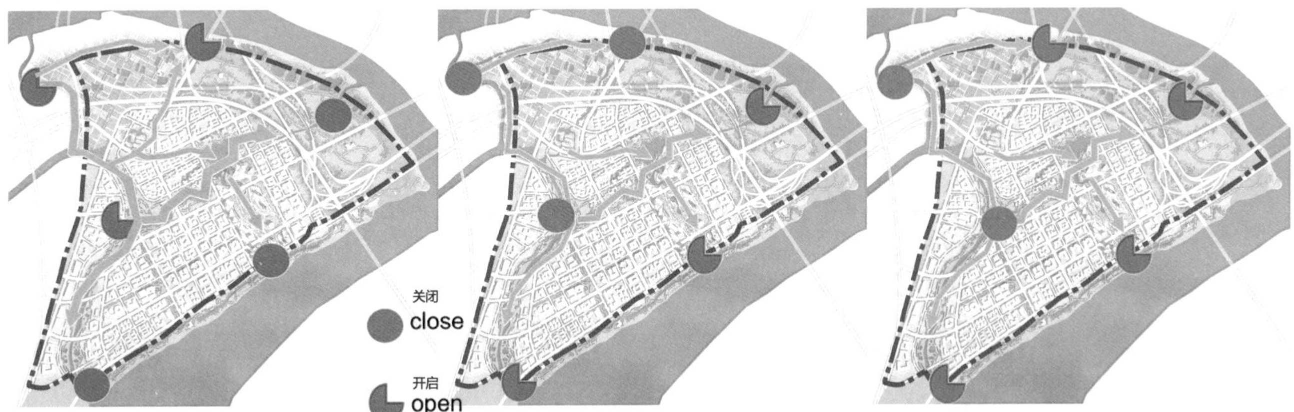

枯水期，开启上游水闸，关闭下游水闸，补给景观和生活用水，对场地内的水进行过滤并实现循环。

During the dry season, open the upstream sluice and close the downstream sluice to replenish the water demand for landscape and life, and filter the site and realize the circulation.

丰水期，开启下游水闸，通过场地内的弹性蓝绿设施涵养水分，多余的水由下游水闸快速排除。

During the rainy season, the downstream sluices are opened to conserve water through the flexible blue-green facilities in the site, and the excess water is quickly drained by the downstream sluices.

暴雨期，开启绝大多数中下游水闸，便于暴雨快速排出，同时利用场地内蓝绿设施和综合排水管网综合排水。

During the rainstorm period, open most of the midstream and downstream sluice gates to allow the rainstorm to be discharged quickly, and at the same time use the blue-green facilities and integrated drainage pipe network in the site for comprehensive drainage.

灰色设施布局——地下水网系统 Gray facility layout—Groundwater network system

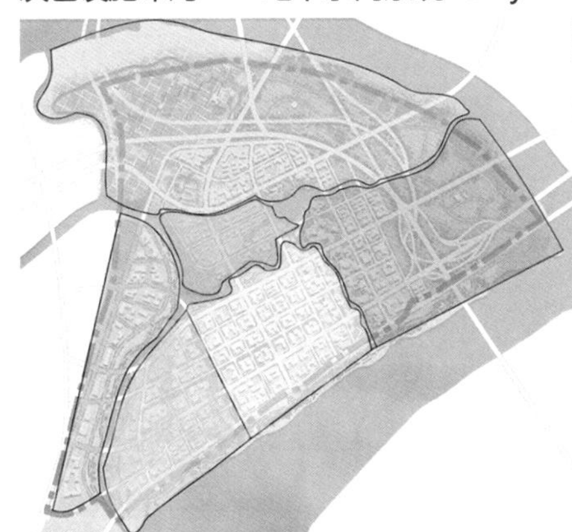

结合 GIS 汇水分析及土地利用现状，将场地地下管网划分为六大汇水区。

Combined with GIS water catchment analysis and land use status, the underground pipe network of the site is divided into six catchment areas.

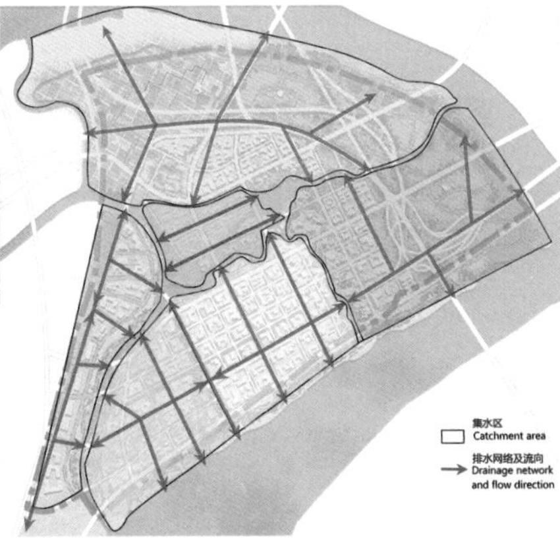

结合地形及道路设置排水网。

Set up drainage network in combination with terrain and road.

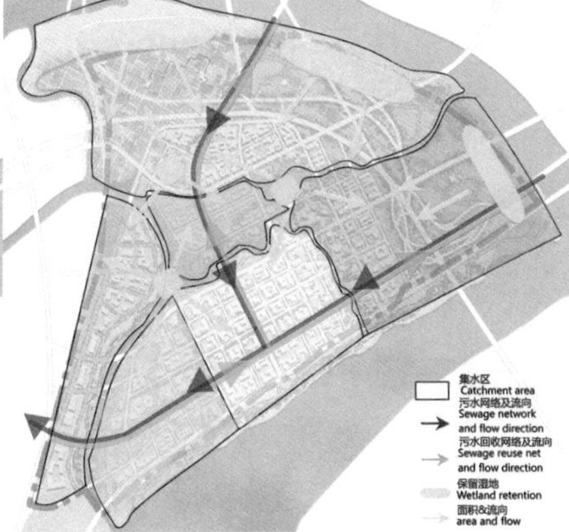

水安全保障＋水环境治理＋水资源利用。

Water security + Water environment treatment + Utilization of water resources.

TOD导向城市组团 Transit-oriented city groups

策略二 功能复合

Strategy 2 Functional composite

TOD环形模式图 TOD unit circle layer pattern map

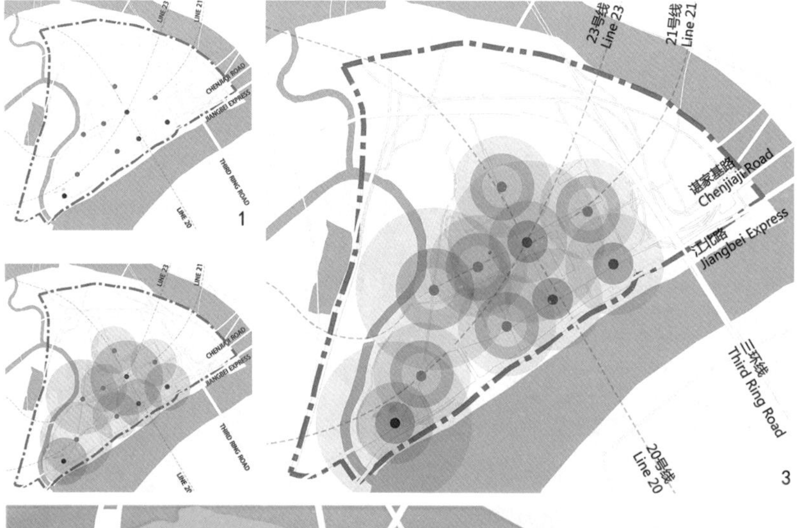

1.根据交通站点确定TOD单元。
Determine TOD units based on traffic stops.
2.形成商贸中心型、居住中心型、公共中心型三类TOD单元。
Types of TOD: business center, residential center and public center.
3.确定TOD单元各圈层开发模式。
Determine the development mode of each circle of the TOD unit.

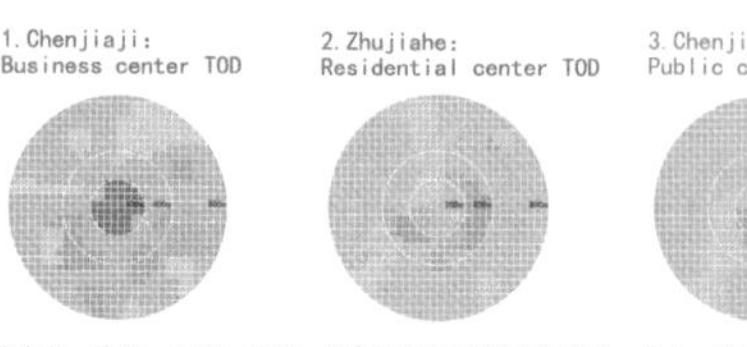

将商业、商务、居住、绿地、服务设施用地整合为核心、中心、边缘圈层，形成多功能TOD单元。
Integrate business, commercial, residential, green space, and service facility land into the core, center, and edge circles to form a multi-function TOD unit.

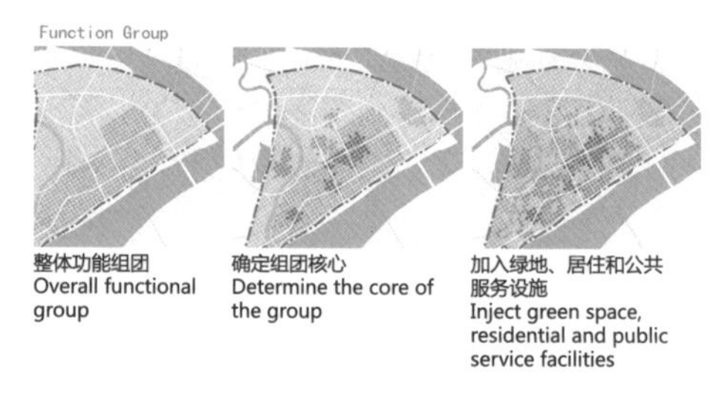

整体功能组团
Overall functional group

确定组团核心
Determine the core of the group

加入绿地、居住和公共服务设施
Inject green space, residential and public service facilities

公共服务设施图 Public service facilities map

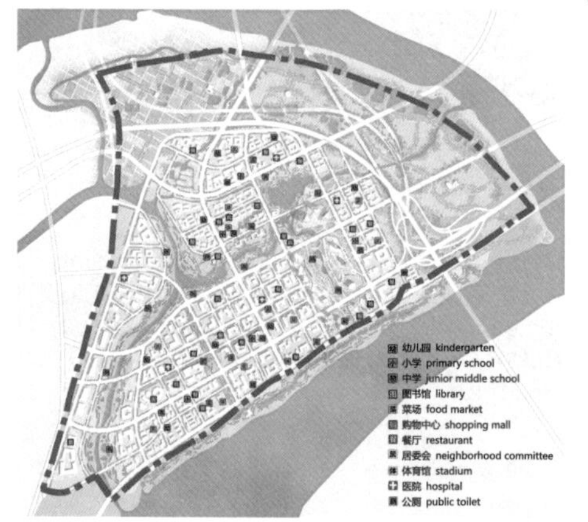

高效交通体系 Efficient transportation system

高效公共交通系统
Efficient public transportation system

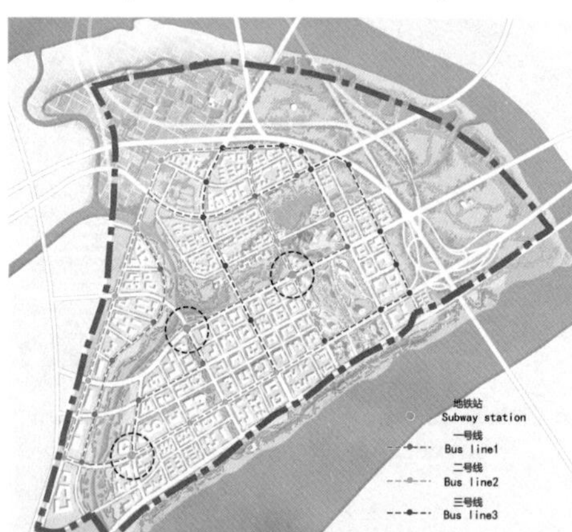

慢性交通系统
Slow-traffic system

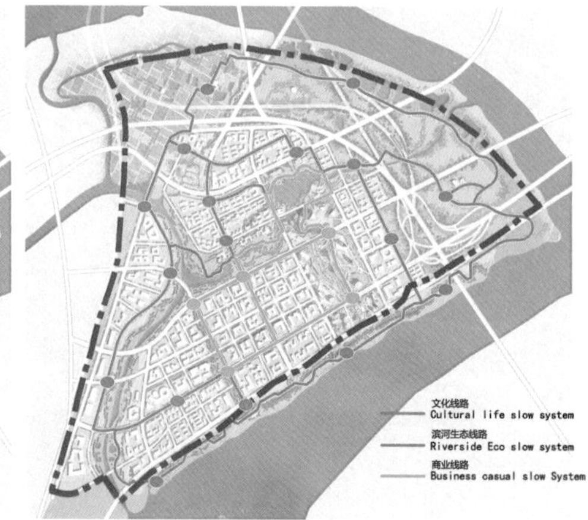

结合慢行绿道布局交通换乘中心，尽量将地面空间释放给市民。

Combine the slow-moving greenway with the layout of the traffic interchange center to free up ground space.

以慢行系统串联活力公共开放空间，尽享场地生态滨水资源，结合商业、医疗、公共服务设施等邻里配置，打造高质量健康城市慢行系统。

Connect the active public open space with the slow-moving system, enjoy the ecological waterfront of the site, and combine the neighborhood configuration such as commercial, medical, and public service facilities to create a high-quality, healthy-city slow-moving system.

公共空间规划 Public space planning

策略三 活力激发
Strategy 3 Vitality

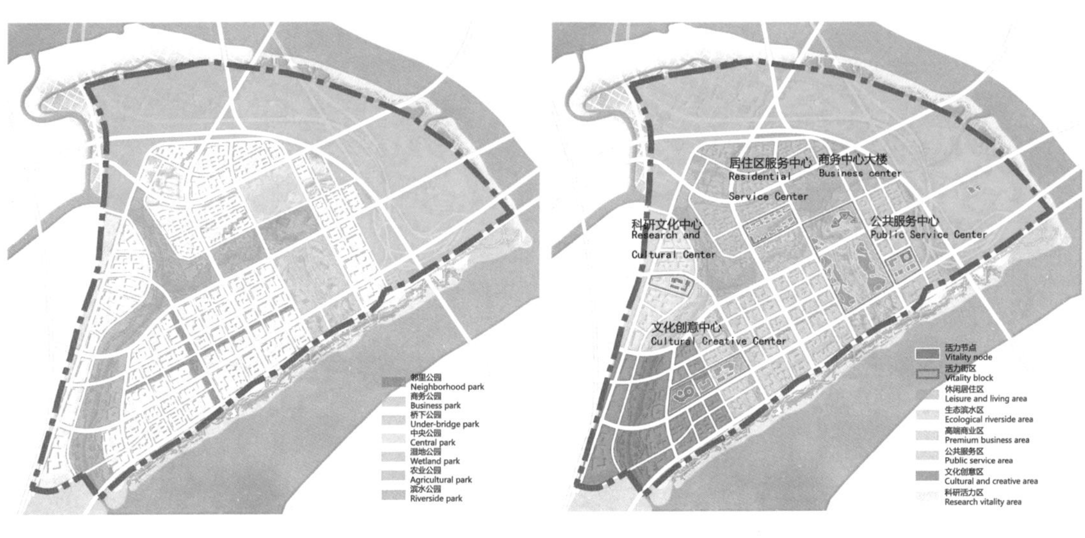

规划各类公园，打造活力街区和核心特色标志建筑，构建充满活力的复合型公共活动空间。

Planning different kinds of parks, creating dynamic blocks and core characteristic sign buildings, and building a dynamic compound public activity space.

城市天际线控制 City skyline control

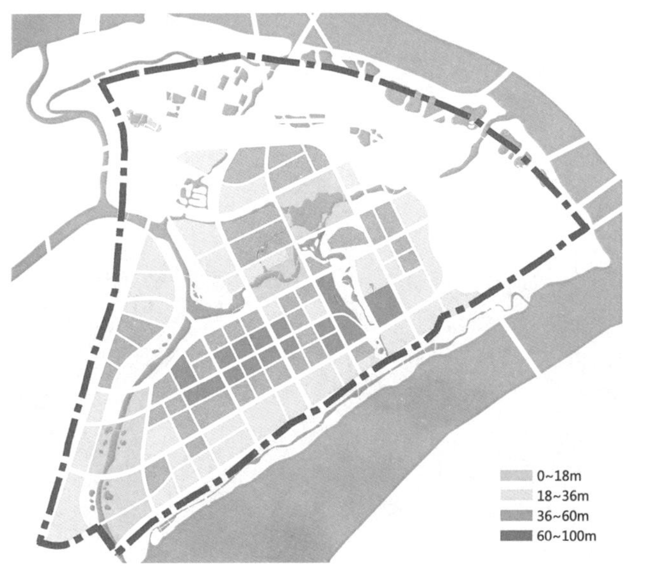

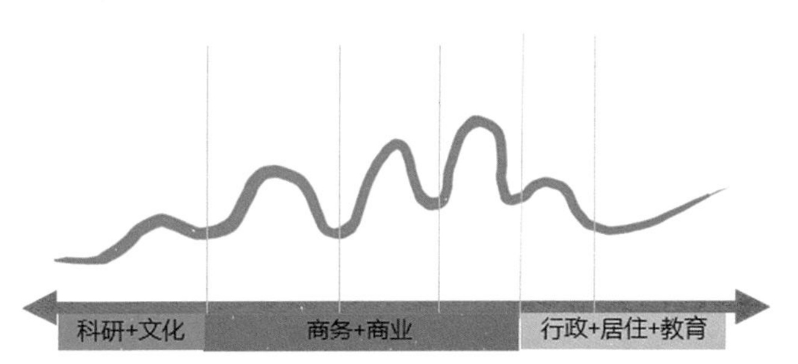

通过控制建筑高度，使其呈现从中部商务区向东、向西下降的趋势，形成独特的城市天际线。

Through the control of the building height, it shows a downward trend from the middle business district to the east and west areas, creating a unique city skyline.

1
4
5
6
7
8
9
10
11
12
13
14
15
16
17
18
19

① 农田湿地 farmland wetland
② 滠水湿地公园 Sheshui River Wetland Park
③ 健康驿站 health station
④ 雨洪公园 rain flood park
⑤ 社区活动中心 community activity center
⑥ 中央公园 central park
⑦ 高端商务中心 high-end business center
⑧ 商务中心绿地 business center green space
⑨ 学校 school
⑩ 滨水商业街 waterfront commercial street
⑪ 朱家河公园 Zhujia River Park
⑫ 科研教育展示基地 exhibition center
⑬ 医院 hospital
⑭ 二七纪念馆 Erqi Memorial Hall
⑮ 博物馆 museum
⑯ 廊带绿地 corridor green space
⑰ 滨江别墅 riverside villa
⑱ 小型商业活动中心 small business center
⑲ 滨江休闲道 riverside leisure road

专项设计 Design of special topic

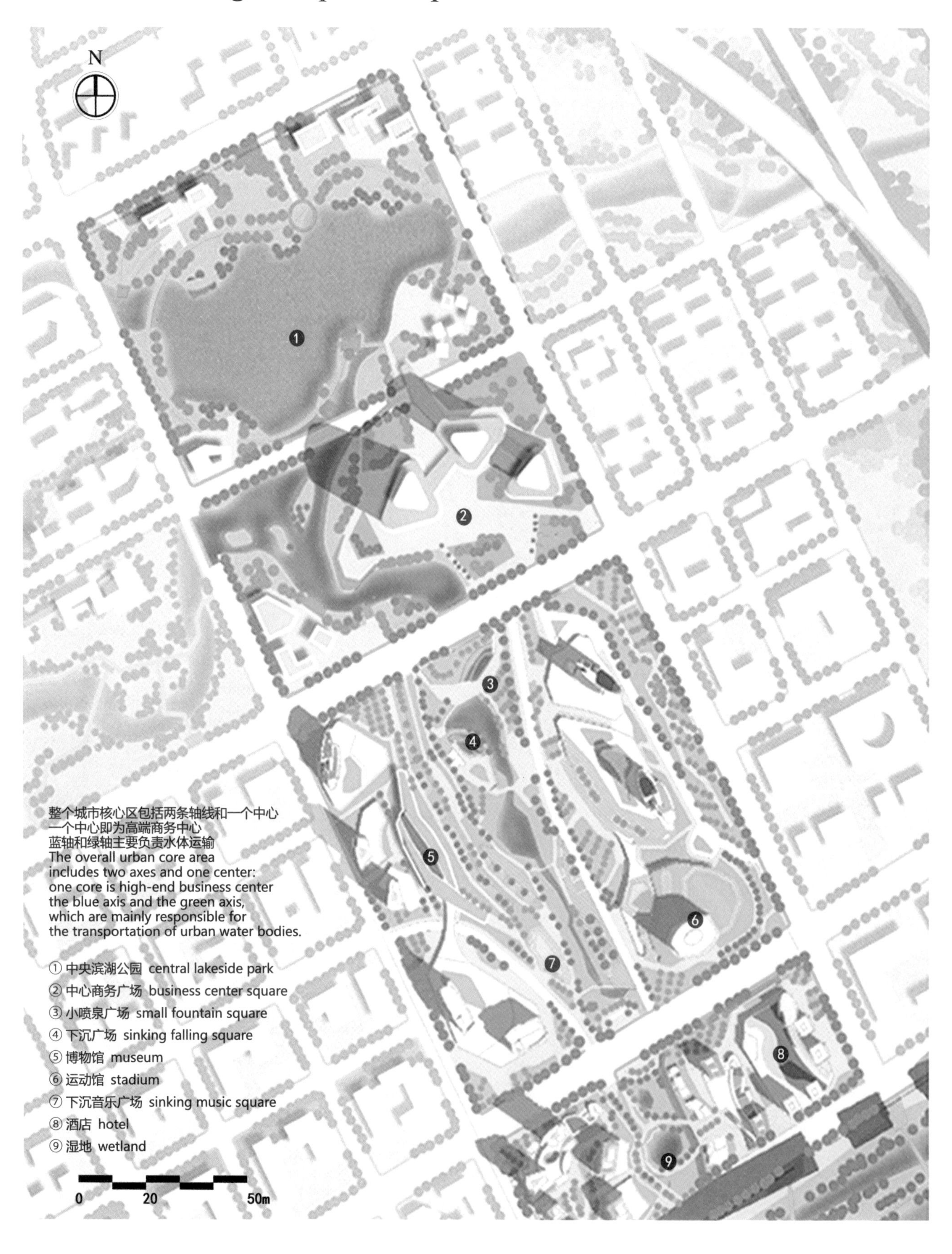

中轴线建筑功能 Central axis building function

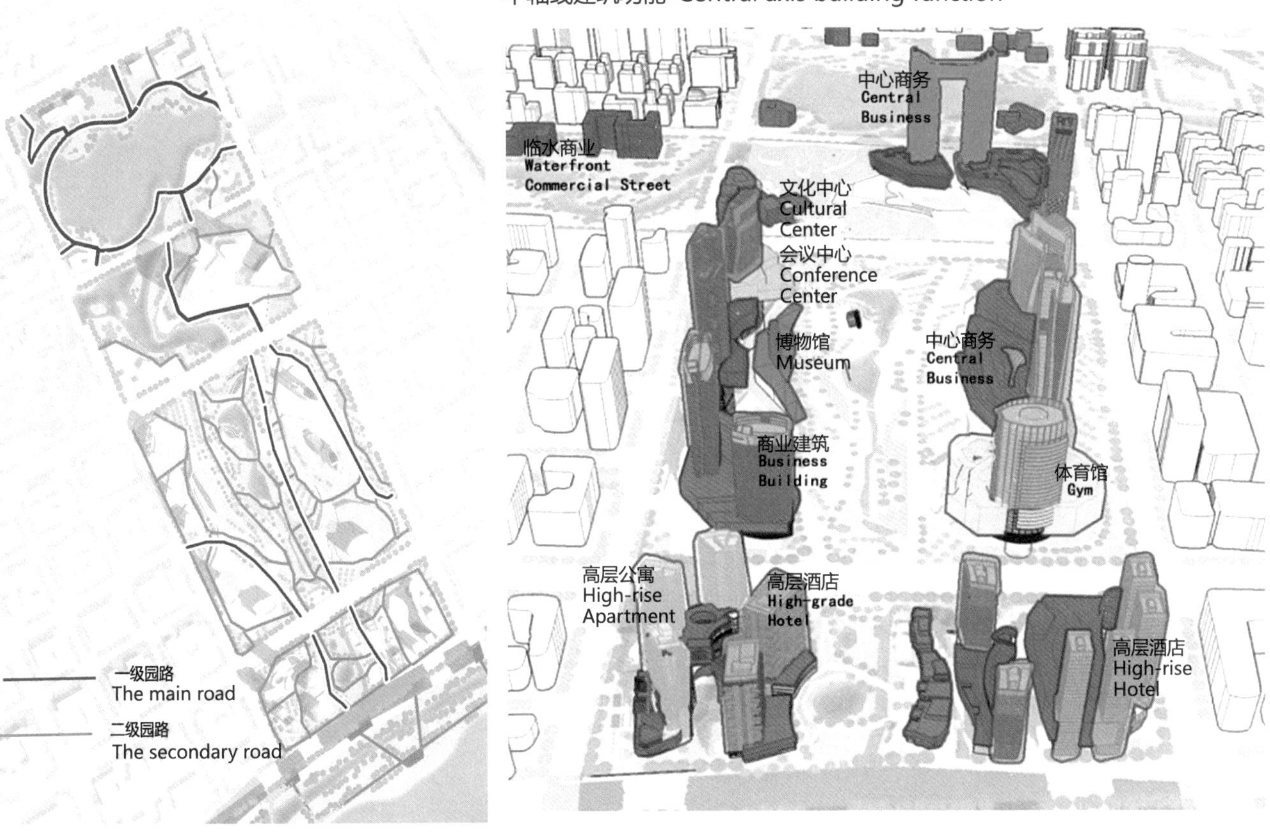

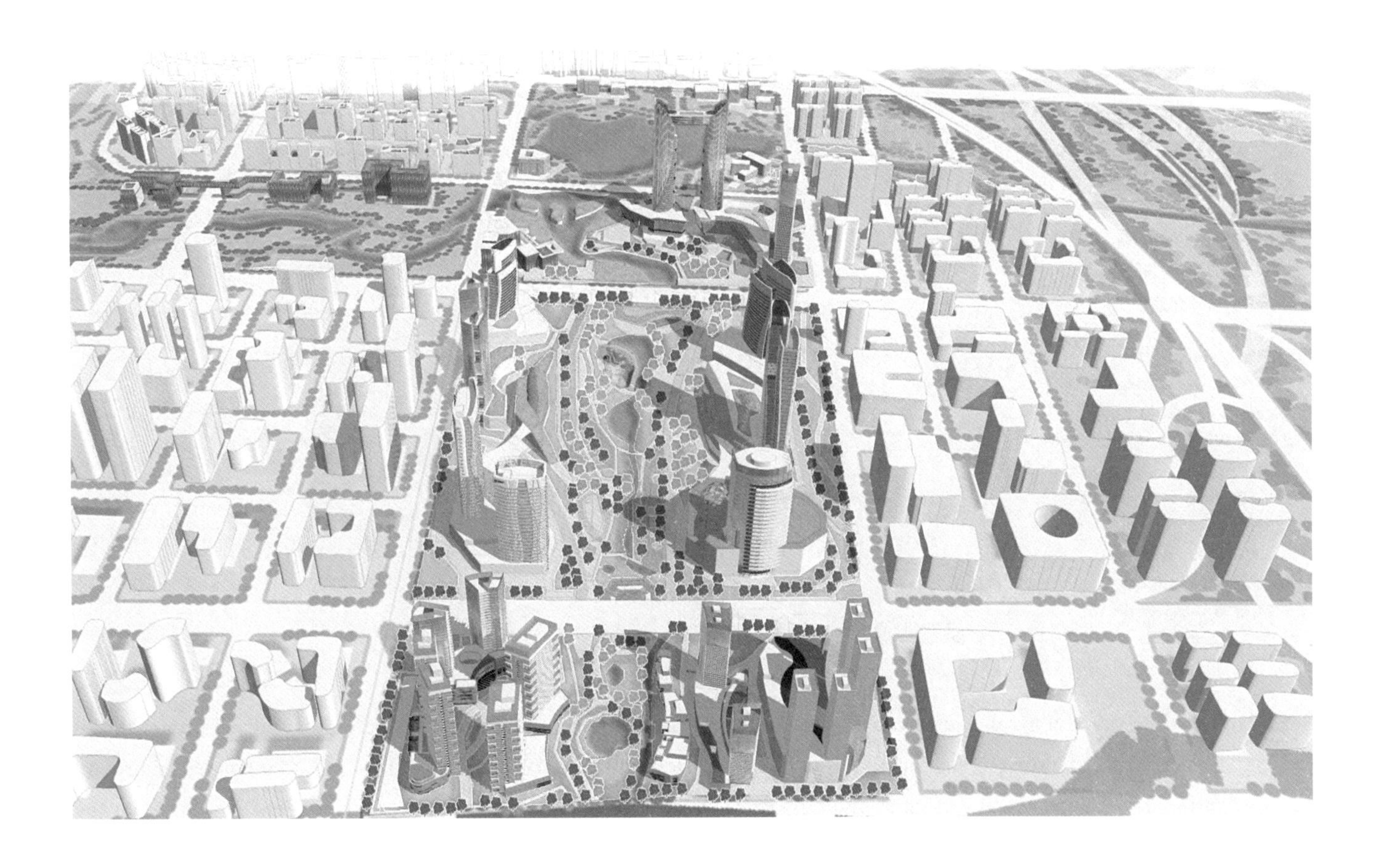

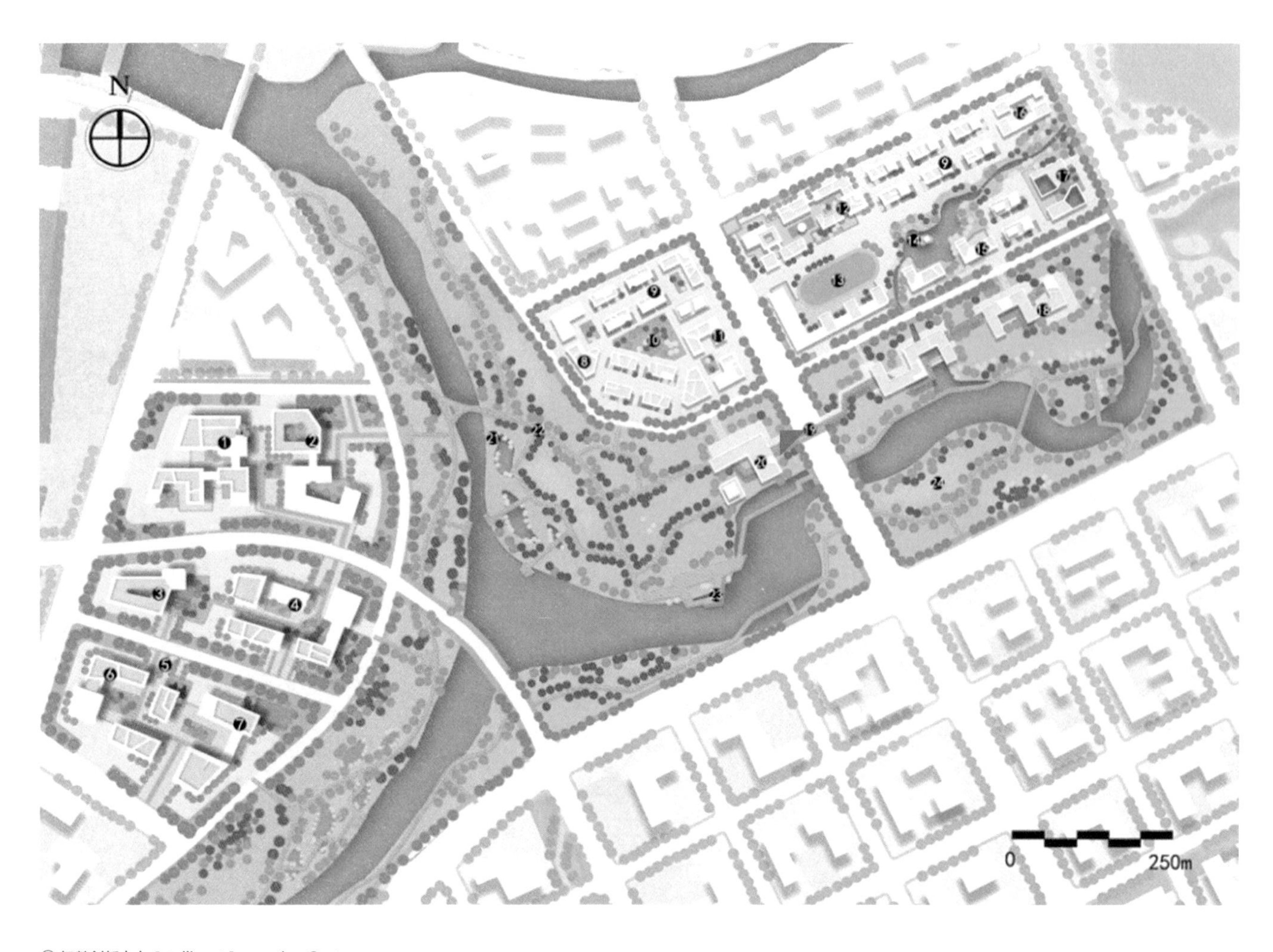

① 智慧创新中心 Intelligent Innovation Center
② 健康科技中心 Health Technology Centre
③ 医院 hospital
④ 医械试验平台 medical apparatus test platform
⑤ 入口广场 entrance plaza
⑥ 人才公寓 Talents Apartment
⑦ 科研机构 scientific research institutions
⑧ 屋顶花园 rooftop garden
⑨ 住宅楼 residential building
⑩ 雨水花园 the rain garden
⑪ 社区健身馆 community gym
⑫ 社区商业中心 community business center
⑬ 学校 school
⑭ 中心花园 Central Garden
⑮ 高级公寓 senior apartment
⑯ 文化展览馆 Cultural Gallery
⑰ 艺术体验中心 Art Experience Center
⑱ 滨水商业街 waterfront commercial
⑲ 空中连廊 street and air corridor
⑳ 美食城 food court
㉑ 人工湿地 artificial wetland
㉒ 滨水步道 waterfront trail
㉓ 景观平台 landscape platform
㉔ 休闲草坡 leisure slope

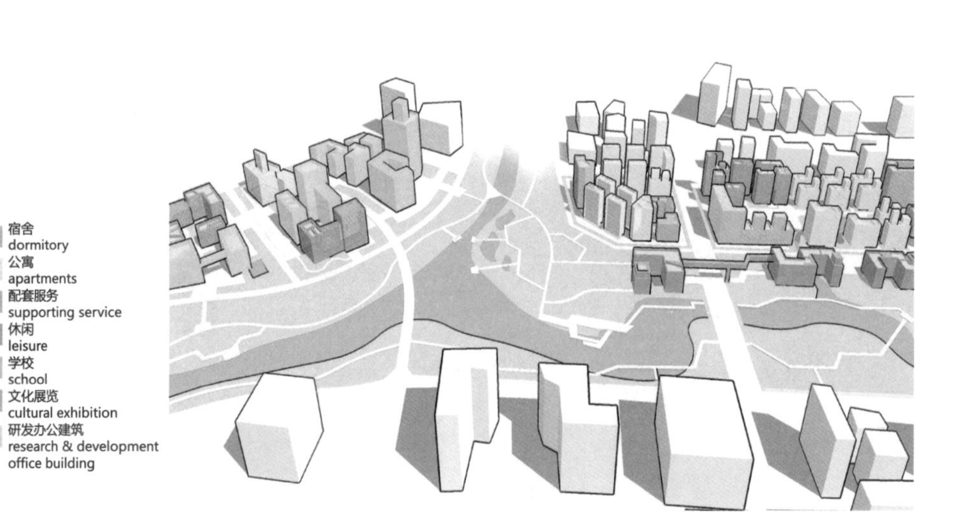

在滨河区域设置较为低矮的餐饮休闲建筑。居住区内设置配套服务点，居住区附近穿插配套学校、文化展览与商业性质的休闲建筑，科研文化区在研发办公建筑中设置附属宿舍作为人才公寓，提供丰富的多样化活力空间。

Set relatively low-rise dining and leisure buildings in the riverside area. Supporting service points will be set up in the residential area, and supporting schools, cultural exhibitions and commercial leisure buildings will be interspersed around the residential area. The scientific research and cultural zone sets up an affiliated dormitory as a talent apartment in the R&D office building to provide a rich and diverse vitality space.

蓝绿设施 Blue green facilities

绿色凹带
concave greenbelt

雨水花园
rain garden

植草沟
the road furrowed with grass

在居住区、广场上设置小块的下凹绿地，在居住区内利用较大的绿地设置雨水花园，营造良好的景观，在道路两边设置植草浅沟，结合城市河道发挥雨水调蓄作用。

Set up small plots of sunken green space in residential areas and squares, and use larger green spaces in residential areas. Set up rainwater gardens on the ground to create a good landscape. Shallow grass-planting ditches are set up on both sides of the road to play the role of rainwater regulation and storage in conjunction with urban rivers.

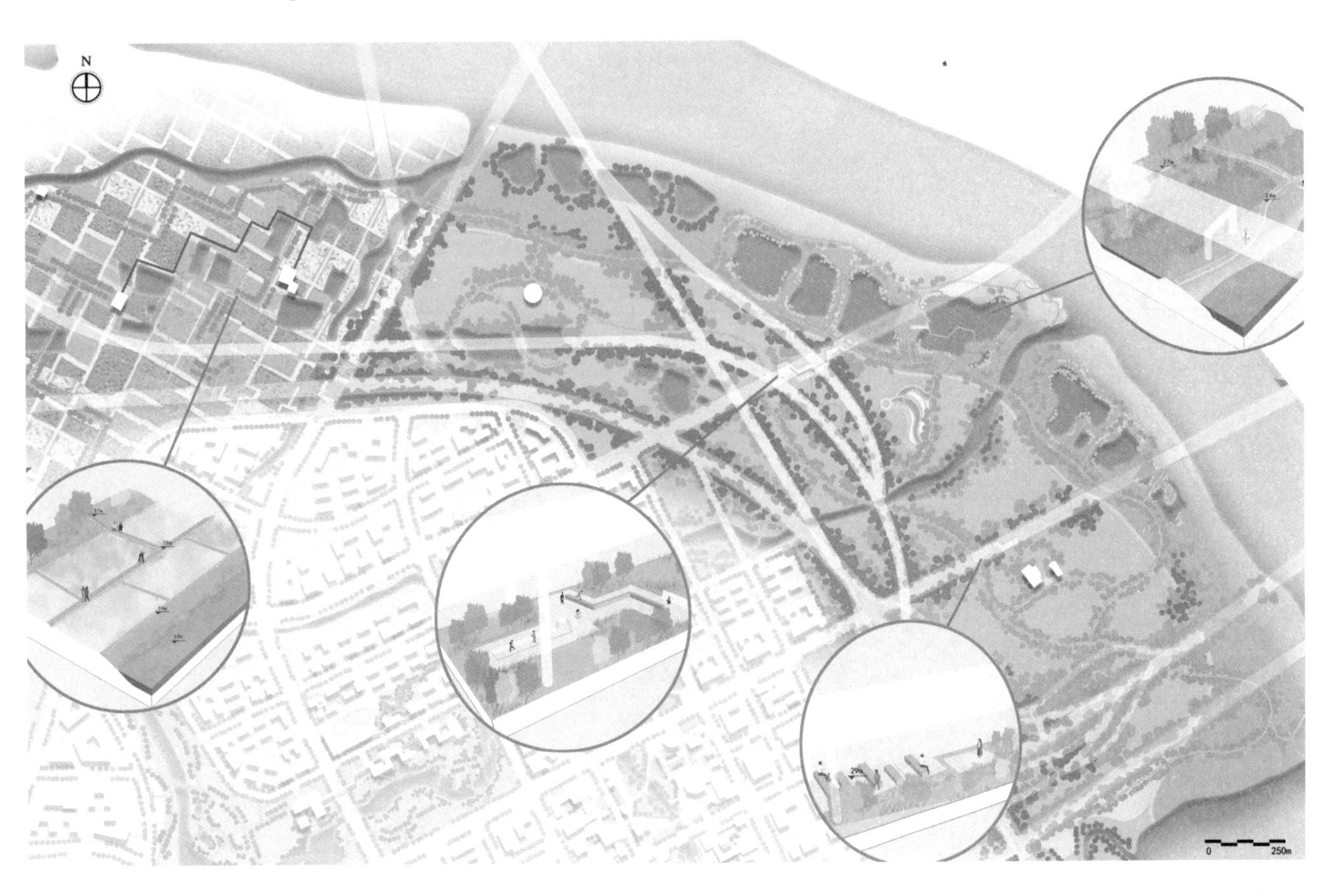

详细设计 Detailed Design

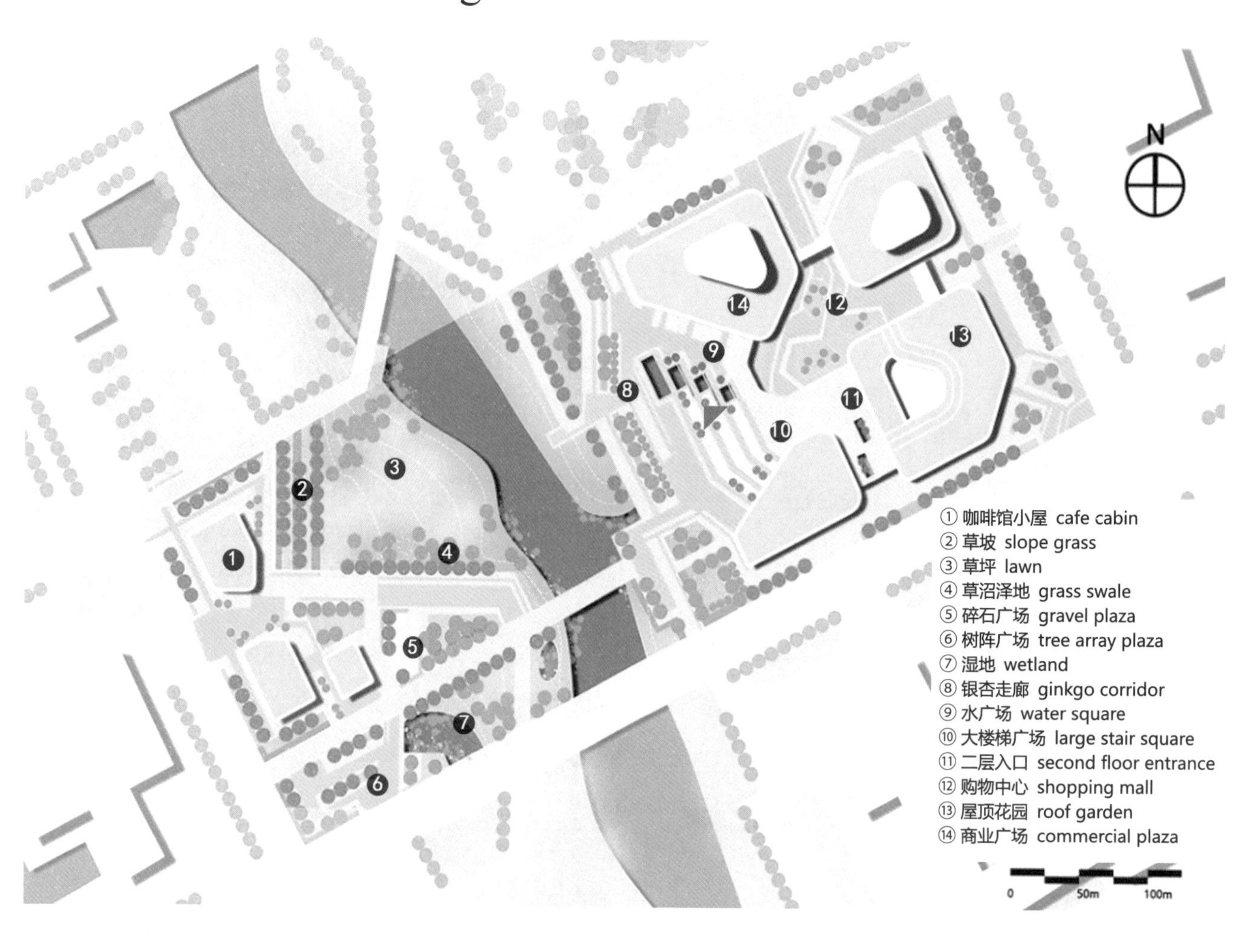

蓝绿设施 Blue & Green Facilities

垂直绿化
vertical planting
绿色屋顶
green roof
湿地
rainwater wetland
水池
pool
沼泽地
grass swale

建筑与景观结构 Building & Landscape Function

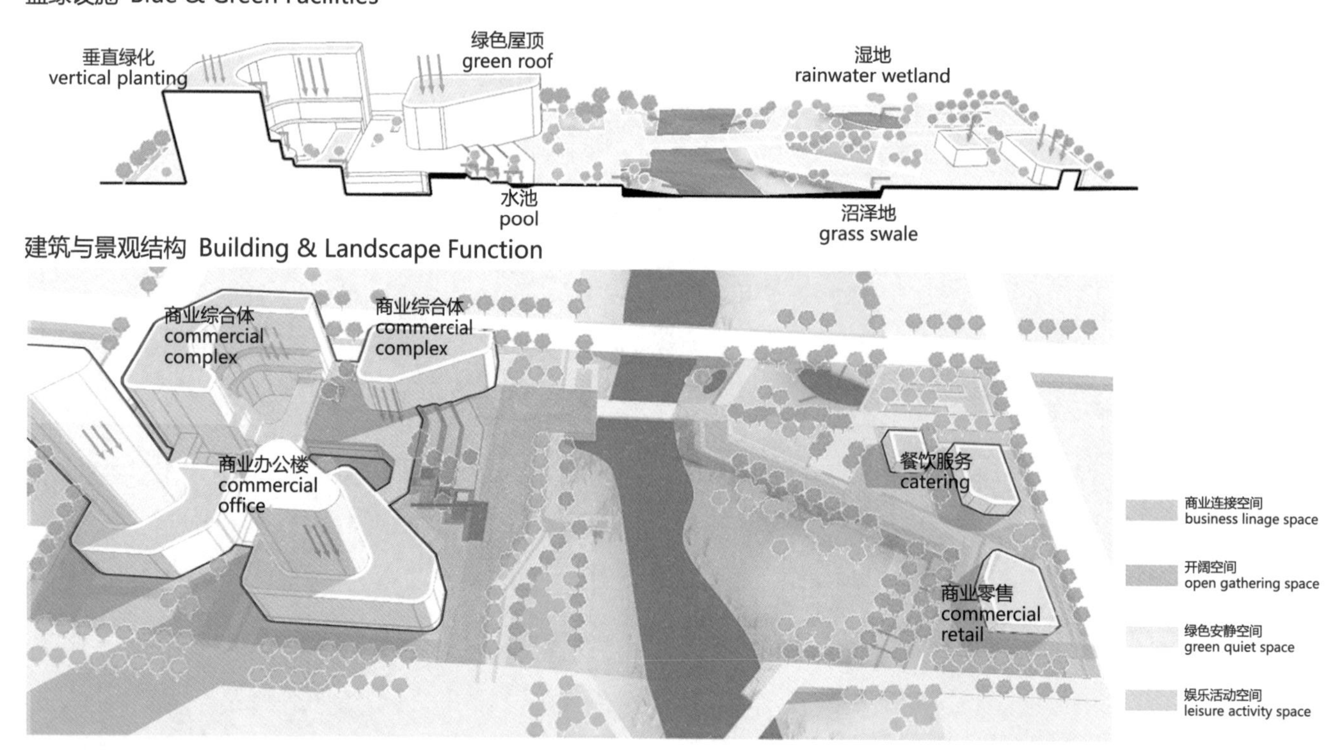

商业裙楼连接高层办公楼和低层商业广场形成商业综合体，低层商业建筑采用退层设计，视线向绿轴内河打开，结合屋顶绿化和立体绿化截留屋面雨水。

The commercial podium connects the high-rise office building and the low-rise commercial plaza to form a commercial complex. The low-rise commercial building adopt a fallback design, open the sight to the river in the green axis, and combine roof greening and three-dimensional greening to intercept roof rainwater.

滨水草坪与综合商业广场隔城市内河相望，是重要的开敞休闲活动空间。商业零售和休闲餐饮建筑组合形成低矮建筑群。由路缘植草浅沟和下凹绿地收集硬质区域雨水。

The waterfront lawn and the comprehensive commercial plaza are separated by the city's inland river. At the same time, they are important open spaces for leisure activities. The combination of commercial retail and casual dining buildings forms a low-rise building group. Rainwater from hard areas is collected by shallow trenches and recessed green spaces along the curb.

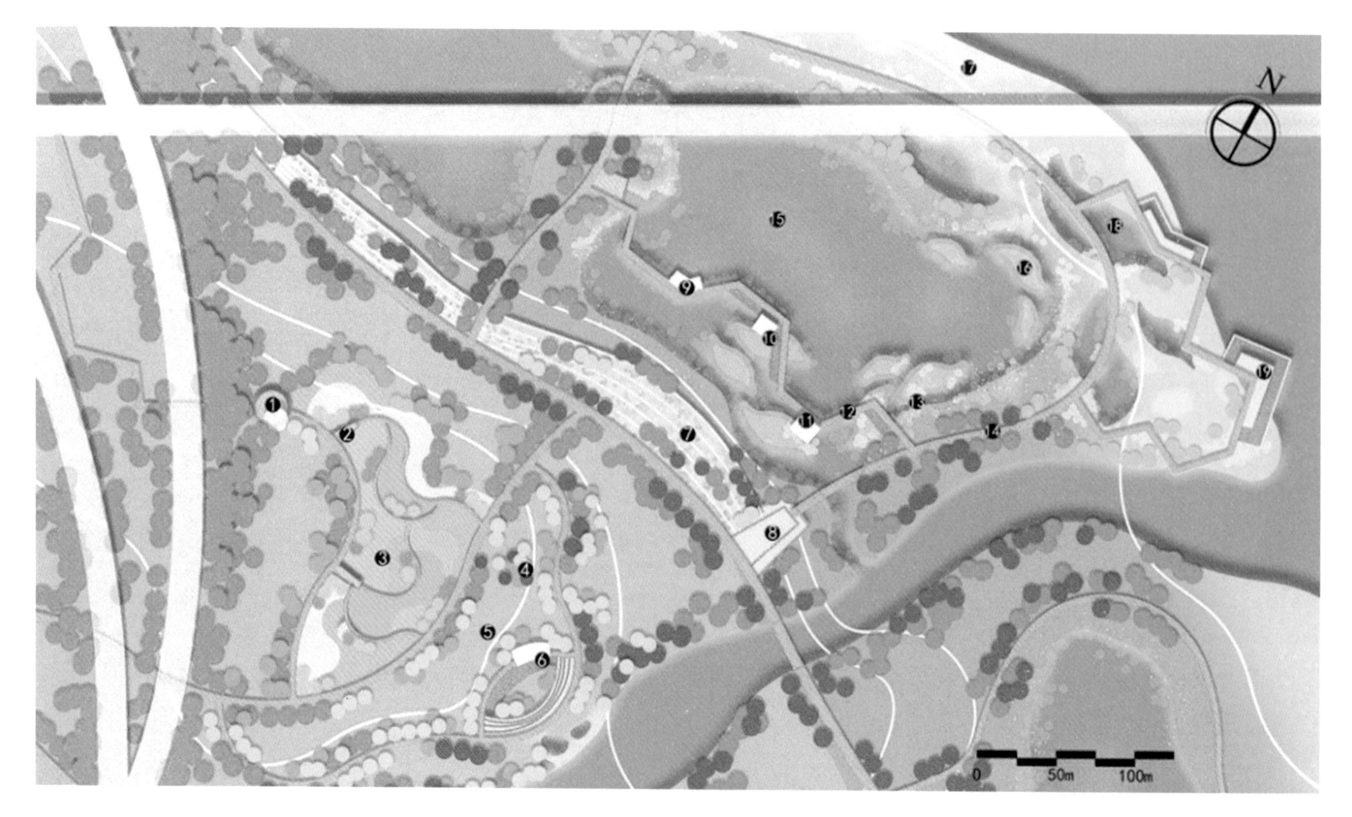

① 集散广场 distribution square
② 观景平台 viewing platform
③ 自然花田 natural flower fields
④ 秋色林 autumn forest
⑤ 阳光草坡 sunny grassy slope
⑥ 休憩亭 rest pavilion
⑦ 生态草堤 ecological grass embankment
⑧ 自然湿地 natural wetland
⑨ 浮水植物盒子 floating plant box
⑩ 沉水植物盒子 submersible plant box
⑪ 挺水植物盒子 uphold plant box
⑫ 湿地探索道 wetland exploration road
⑬ 芦花荡 reed-pond
⑭ 湿地栈道 wetland plank road
⑮ 自然湿地 natural wetland
⑯ 生态小浮岛 ecological small floating island
⑰ 季节性滩涂 seasonal tidal flat
⑱ 湿地小塘 wetland pond
⑲ 滨水平台 waterfront platform

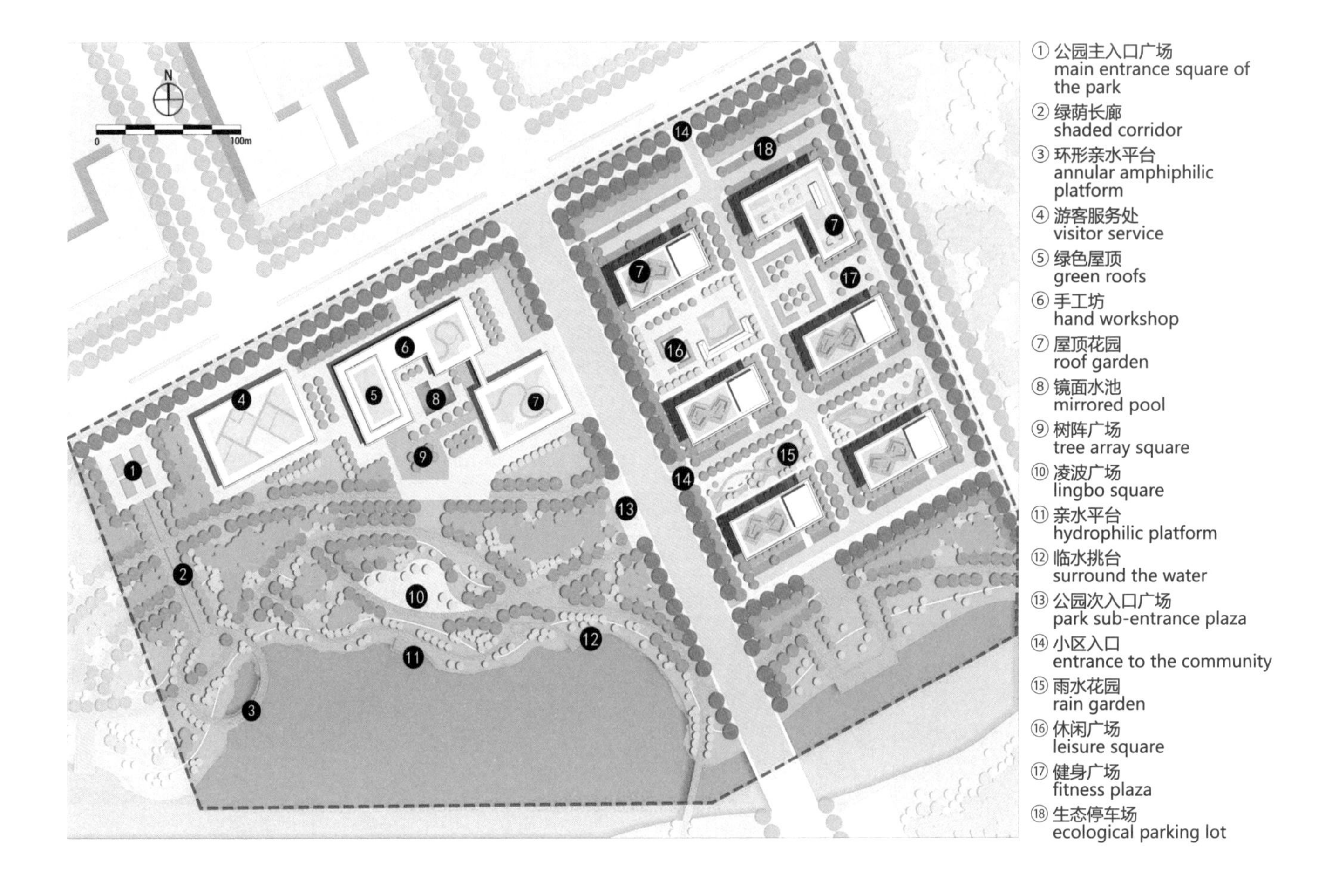

公园滨湖区域整体环境较为自然，普遍采用缓坡入水的形式，设置少量亲水平台和栈道，并种植各种湿地植物。

The overall environment of the lakeside area of the park is relatively natural, and the form of gentle slope into the water is generally adopted, setting up a small number of hydrophilic platforms and plank roads, and planting various wetland plants.

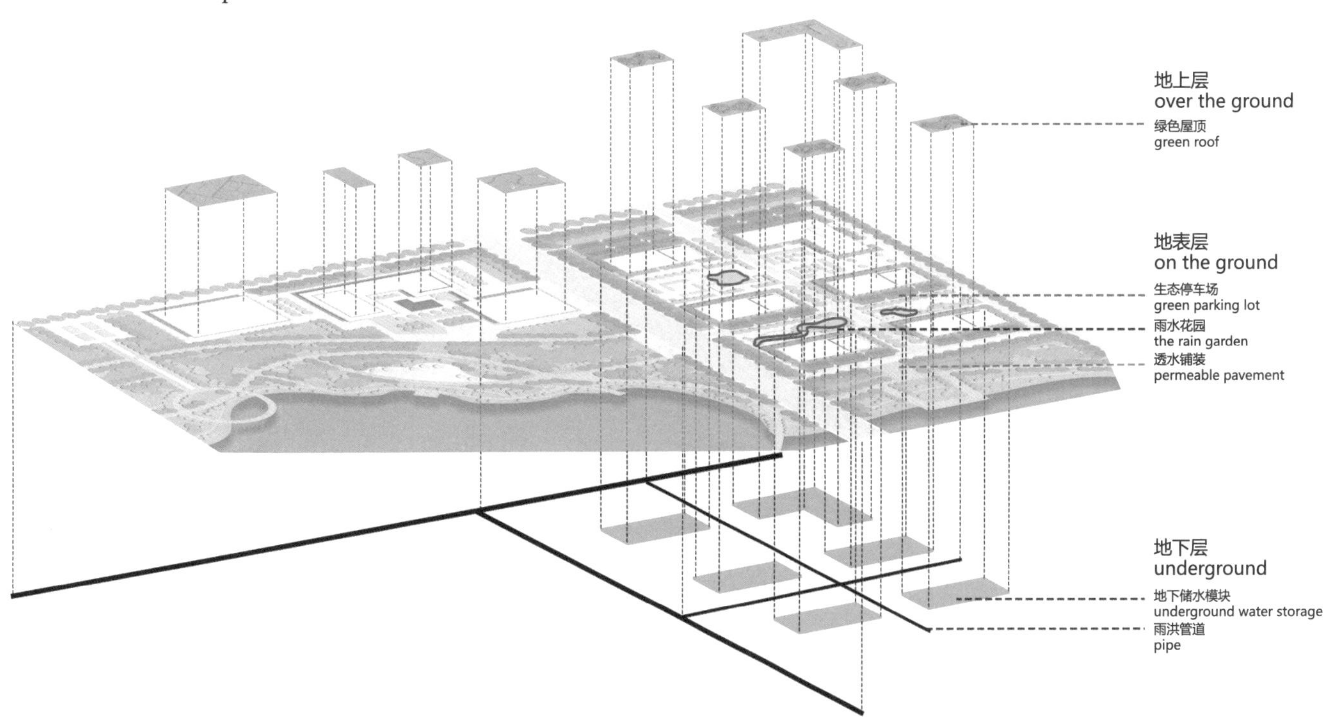

居住小区建筑设置有屋顶花园，地面空间以透水铺装为主，并设有生态停车场和雨水花园，在建筑地下设有雨水收集模块，以便收集和利用雨水。

The residential building is equipped with a roof garden, with permeable pavement as the main ground space, setting ecological parking lots and rainwater gardens. And rainwater collection modules are installed under the building to collect and utilize rainwater.

无人驾驶之城

Driverless

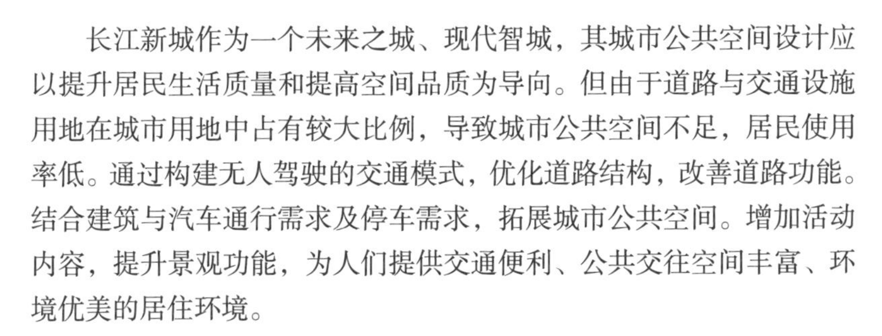

长江新城作为一个未来之城、现代智城，其城市公共空间设计应以提升居民生活质量和提高空间品质为导向。但由于道路与交通设施用地在城市用地中占有较大比例，导致城市公共空间不足，居民使用率低。通过构建无人驾驶的交通模式，优化道路结构，改善道路功能。结合建筑与汽车通行需求及停车需求，拓展城市公共空间。增加活动内容，提升景观功能，为人们提供交通便利、公共交往空间丰富、环境优美的居住环境。

As a city of the future and a modern smart city, the design of urban public space in Yangtze River New City should be oriented to improve the quality of life of residents and enhance the quality of space. However, as the land for roads and transportation facilities occupies a large proportion of the urban land, the urban public space is insufficient and has low utilization rate by residents. By constructing an unmanned transportation mode, the road structure is optimized and the road function is improved. Combine the building and car passage demand and parking demand to expand urban public space. Increase the content of activities and improve the landscape function to provide people with convenient transportation, abundant public interaction space and beautiful living environment.

岑清雅
CEN Qingya

李娅琪
LI Yaqi

马薛骑
MA Xueqi

肖乾坤
XIAO Qiankun

石伟
SHI Wei

向炀
XIANG Yang

叶阳
YE Yang

设计策略 Design Strategy

城市空间结构的演变 Evolution of Urban Spatial Structure

高密度路网结构 high density road network structure

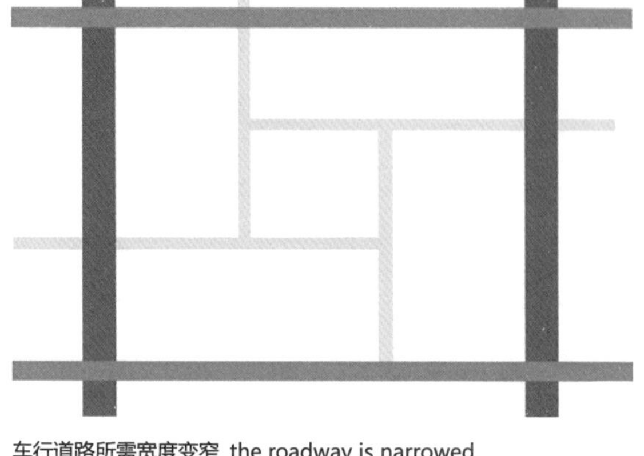

车行道路所需宽度变窄 the roadway is narrowed

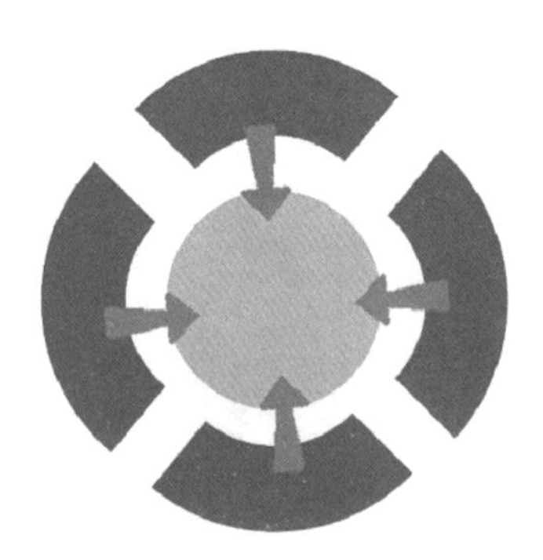

原单核单中心模式 one core

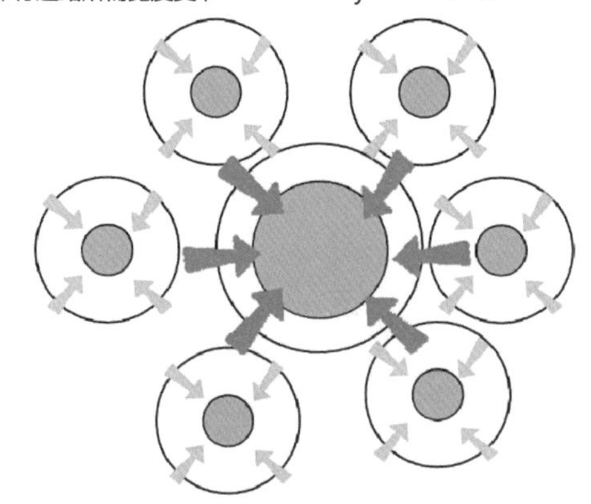

多核化模式，片区综合体 multiple cores

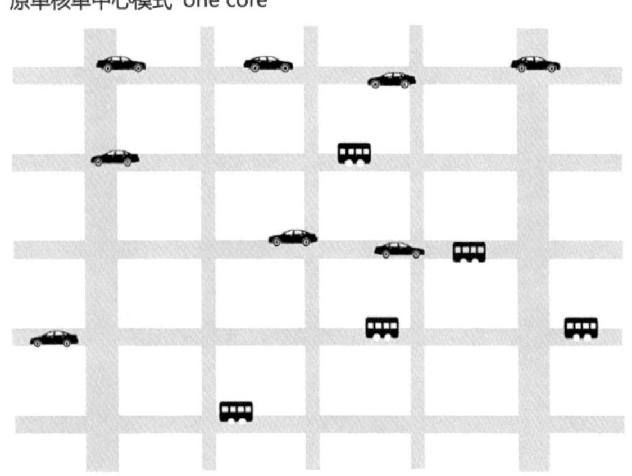

传统城市道路分级，车行为主 traditional urban road

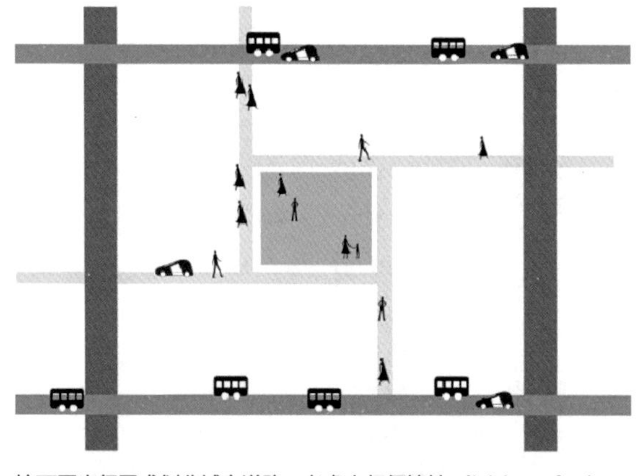

按不同出行需求划分城市道路，考虑人行便捷性 division of urban roads

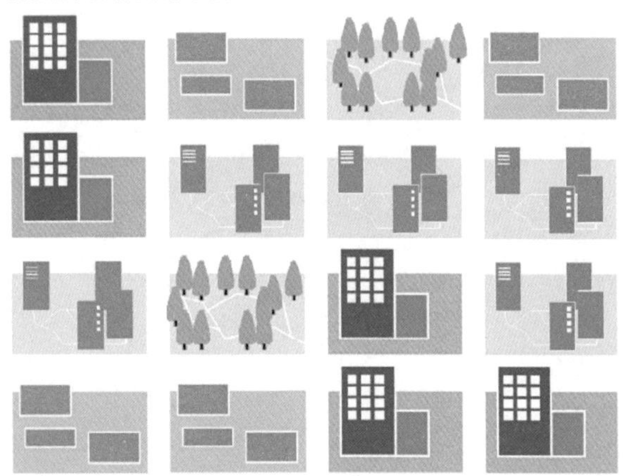

为了提高通行便捷性将用地划分为小街区 small blocks

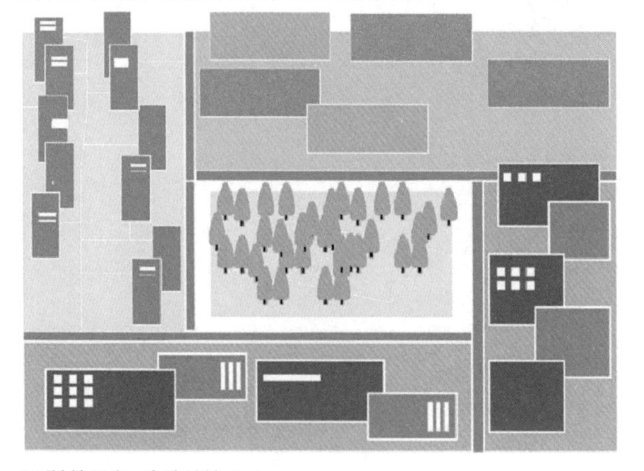

L形地块围合，有连续性 L shape

渗透于公共空间的道路 Road Penetrating Into Public Space Structure

车行道路所需宽度变窄，密度降低 the required width of the road is narrowed and the density is reduced

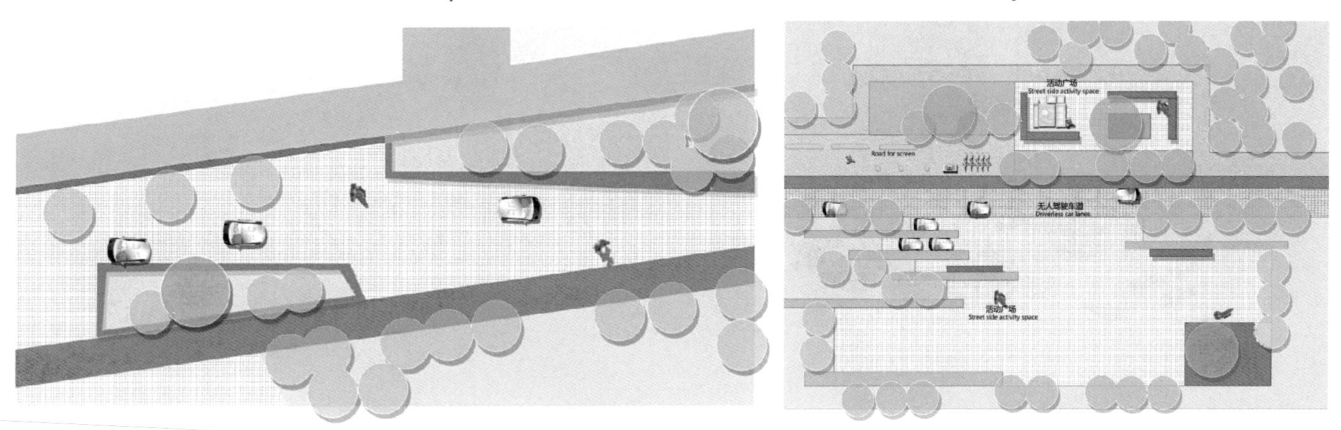

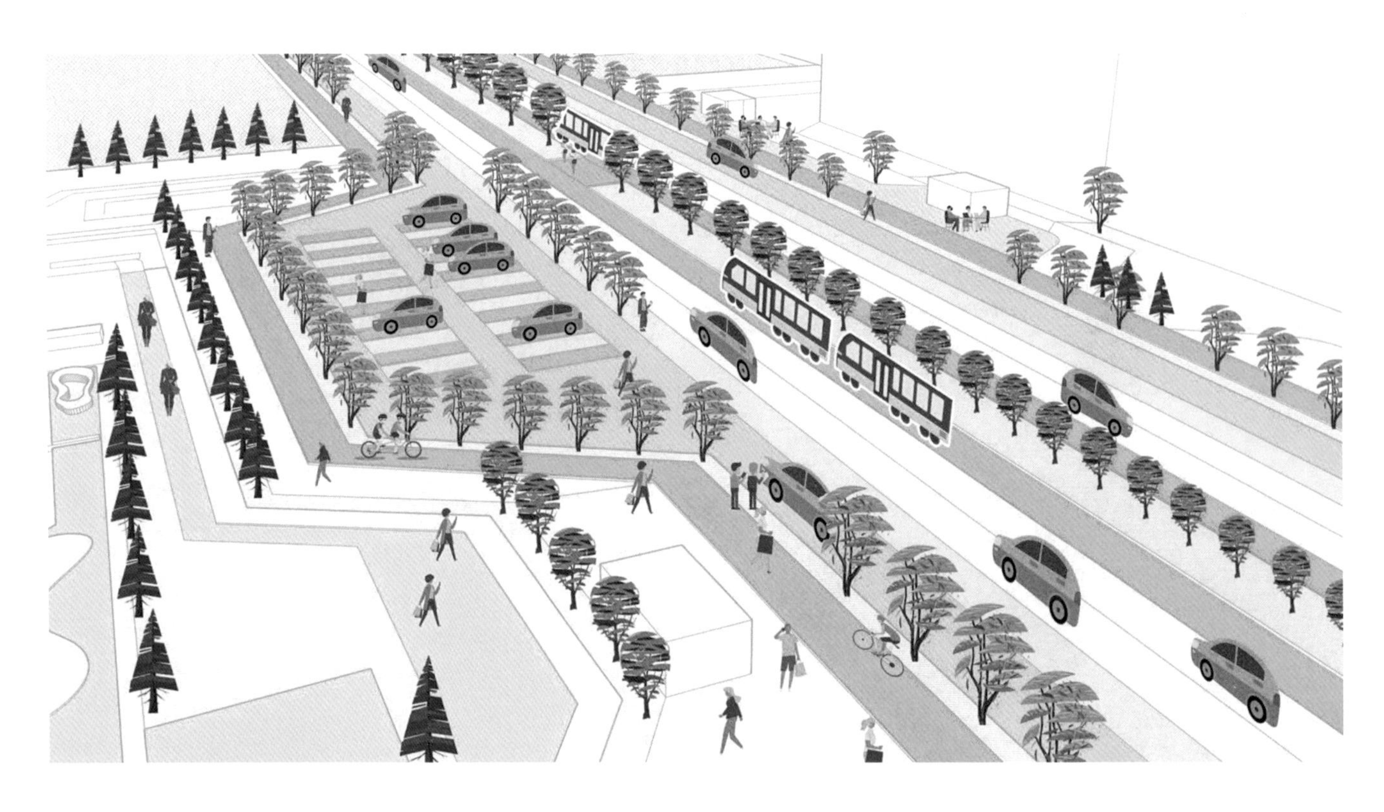

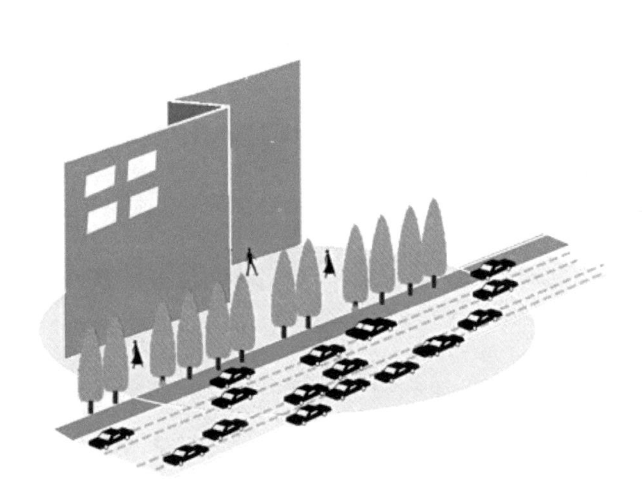

封闭避让 closed

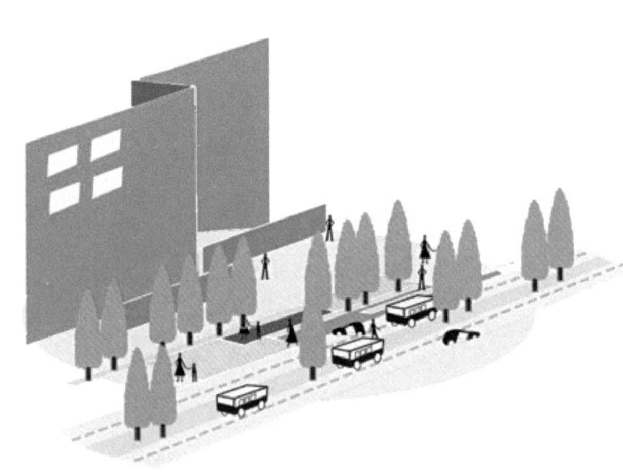

多元融合创造经济效益 diversified integration creates economic benefits

停车空间分散 scattered parking space

停车空间集中 concentrated parking space

商务服务公共空间 Business Service Public

传统广场与道路直接衔接
the traditional square is directly connected to the road

科技展览，创造城市标志
science and technology exhibitions, creating city symbols

传统集散广场 traditional distribution square

垂直空间、景观空间成为建筑功能在外的延伸
the vertical space, the landscape space becomes the extension of the building function outside

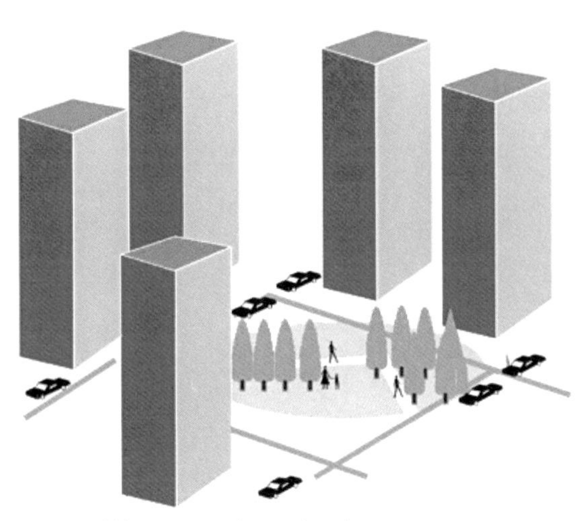

单一景观功能 single landscape function

多功能片区核心 multi-function area core

封闭式小区 gated community

半开放小区 semi-open community

N
100
400
200
600m
1
14
3
6
13
2
10
12
4
7
9
10
6
6
11
8
5
15

① 府河湿地 Fu River Wetland

② 四季公园 Four Seasons Park

③ 谌家矶中学 Chenjiaji Middle School

④ 体育中心 Sports Center

⑤ 朱家河滨江公园 Zhujia Riverside Park

⑥ 城市集中停车楼 urban centralized parking building

⑦ 四季渠绿带 Sijiqu Green Belt

⑧ 江滩公园 River Beach Park

⑨ 中心商业 central business

⑩ 国际金融中心 International Financial Center

⑪ 江城湾 Jiang cheng Bay

⑫ 谌家矶中心医院 Chenjiaji Central Hospital

⑬ 图书馆 Library

⑭ 观鸟露营公园 Bird Watching Camping Park

⑮ 滩涂栈桥 tidal flat pontoon

⑯ 湿地管理处 wetland management division

专项设计 Design of Special Topic

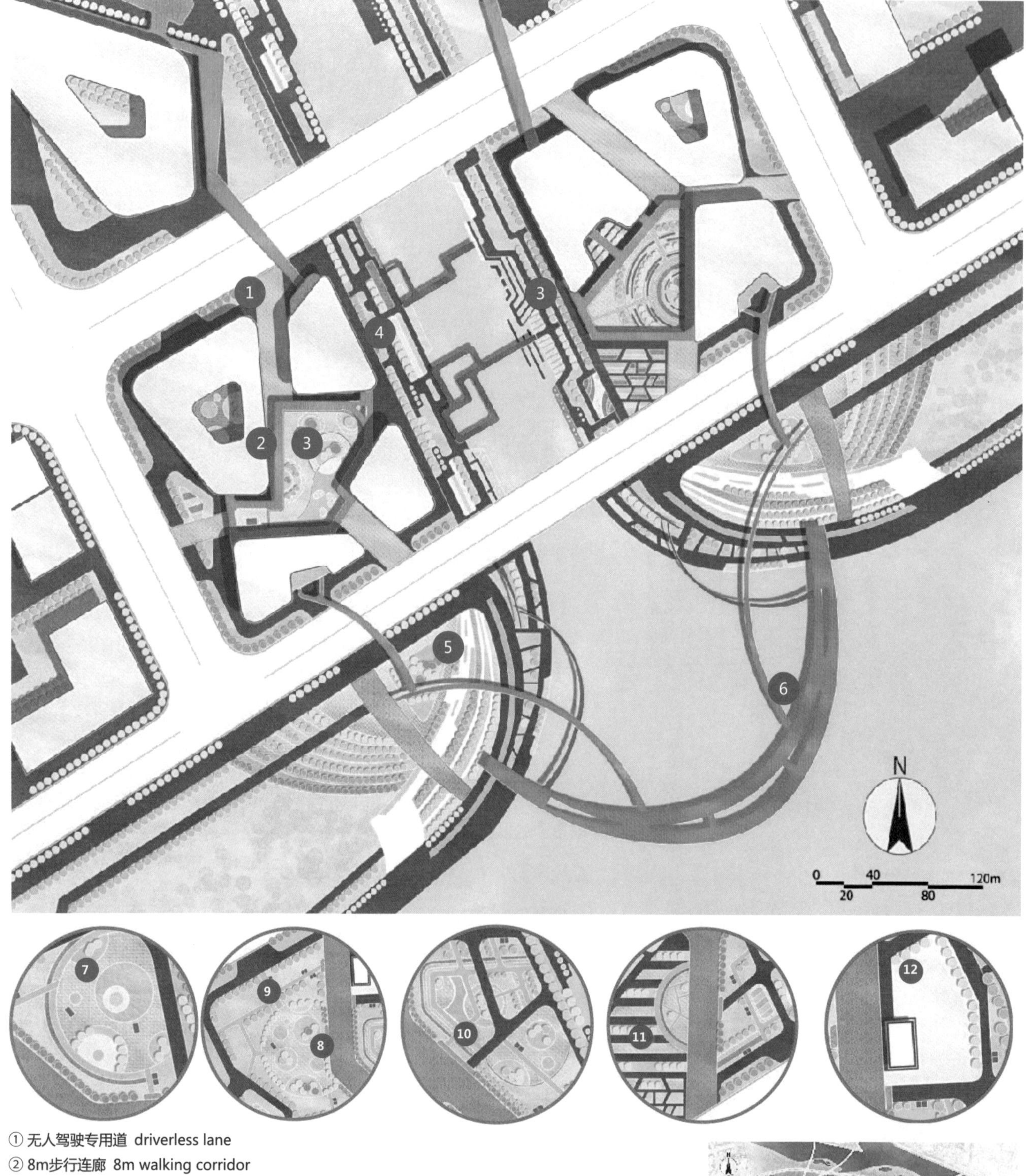

① 无人驾驶专用道 driverless lane
② 8m步行连廊 8m walking corridor
③ 商务区二层公共活动空间 public activity space on the second floor of the business district
④ 滨江绿带 riverside greenbelt
⑤ 江滩公园 river beach park
⑥ 江滩廊桥 river beach covered bridges
⑦ 底层休闲活动场地 ground floor leisure activity space
⑧ 无人驾驶汽车停车点 driverless car parking spots
⑨ 建筑电梯 construction elevator
⑩ 人行步道 pedestrian space
⑪ 景观花带 landscape flower belt
⑫ 底层无人驾驶汽车停车场 driverless car parking on ground floor

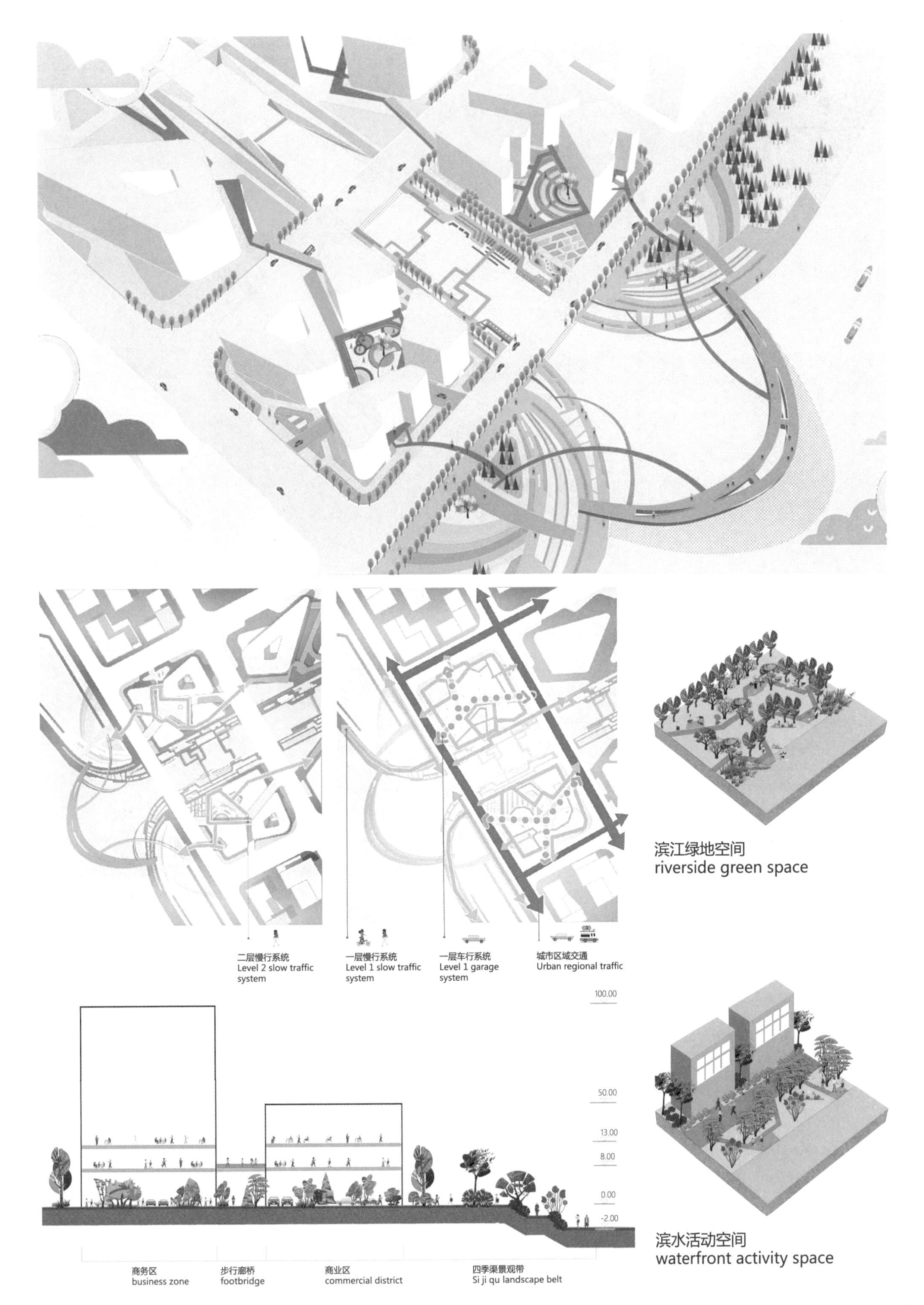

滨江绿地空间
riverside green space

滨水活动空间
waterfront activity space

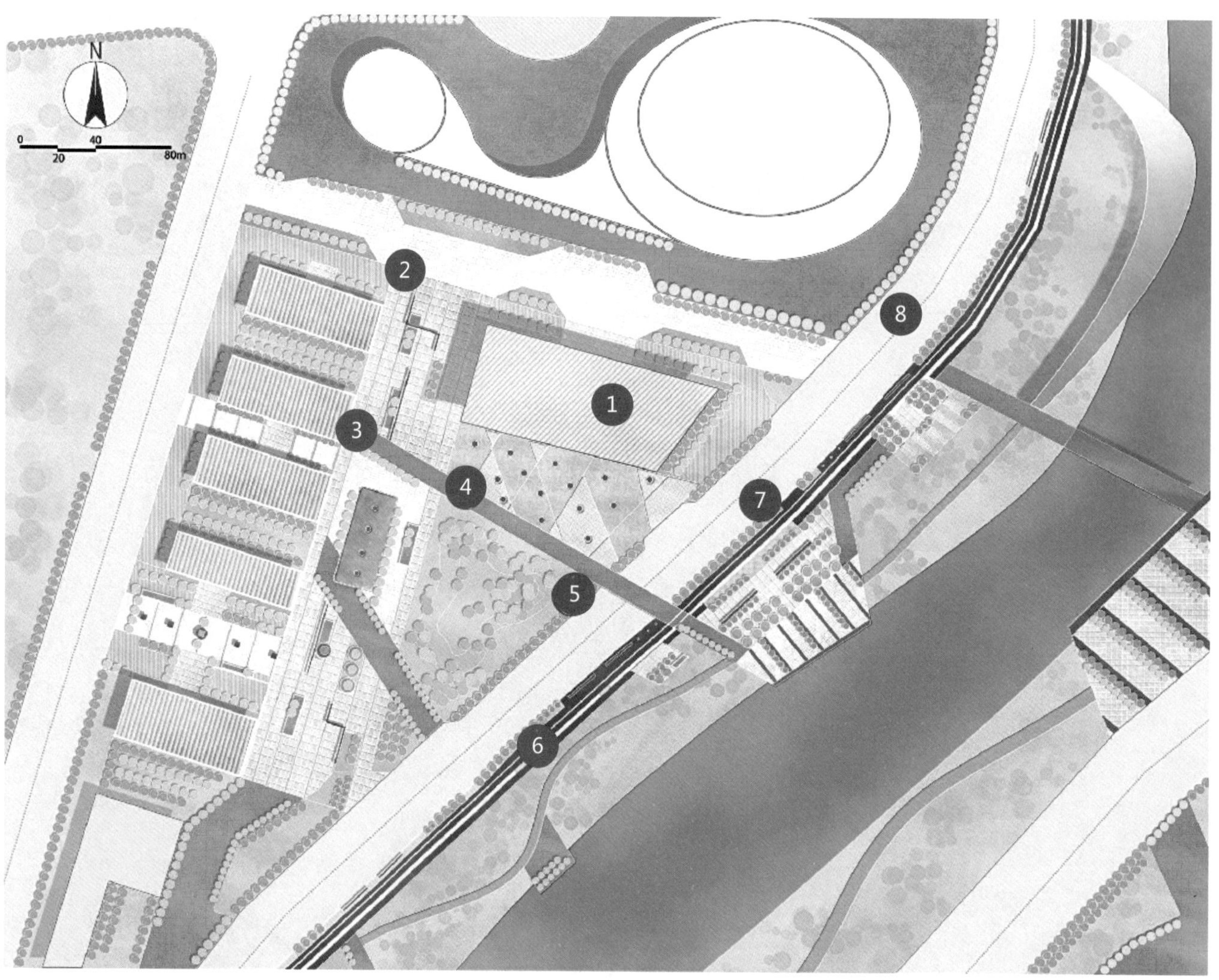

① 科技广场 science square
② 商业广场 business square
③ 景观喷泉 landscape fountain
④ 节庆草坪 festival lawn
⑤ 无人驾驶观光车 driverless sightseeing car
⑥ 滨水活动平台 waterfront activity platform
⑦ 朱家河观景台 Zhujia River viewing platform
⑧ 城市文明标志 signs of urban civilization

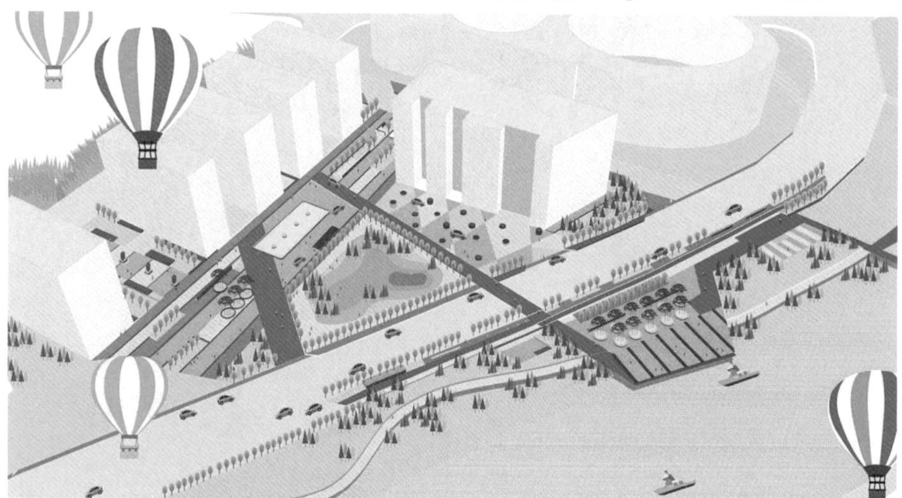

科技展示空间
science and technology exhibition space

城市文化展示空间
urban civilization exhibition space

将释放的街旁空间用于无人驾驶科技的展示以及 VR、5G 等新兴科技的感知体验，并沿无人驾驶旅游线路展示城市文明，为人们提供商业休闲、科技展示与城市文化宣传等服务。

The released roadside space is used for display of driverless technology, VR, 5G and other emerging technology perception experience. At the same time, urban civilization is displayed along the driverless tourism route to provide people with business leisure, science and technology display and urban culture publicity services.

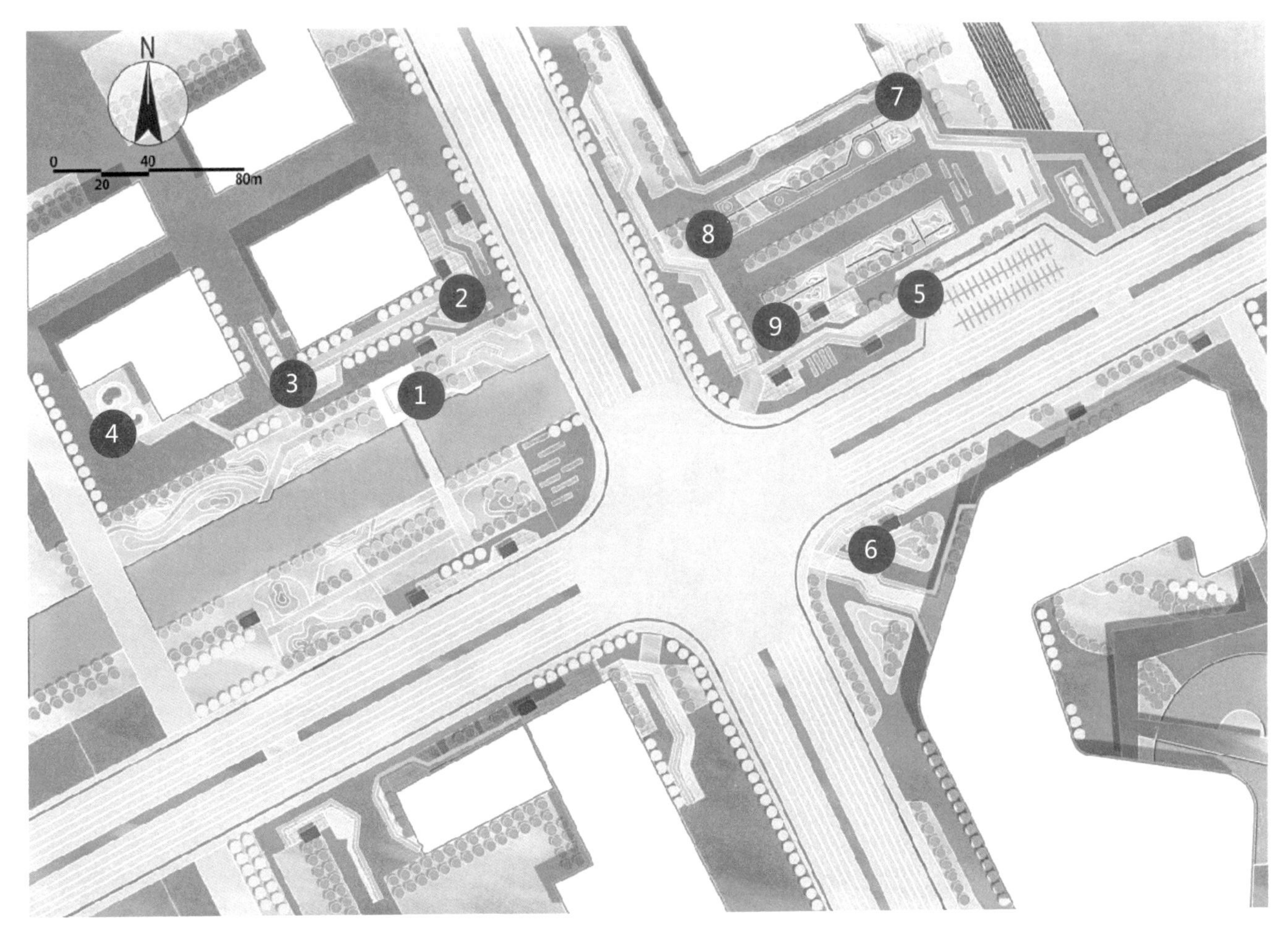

① 城市渠道 urban channel
② 滨水公园 waterfront park
③ 街旁游园 community park
④ BRT通道 BRT channel
⑤ 应急停车空间 emergency parking space
⑥ 商业建筑 commercial buildings
⑦ 滨水广场 waterfront plaza
⑧ 商业空间 business space
⑨ 业态场地 format site

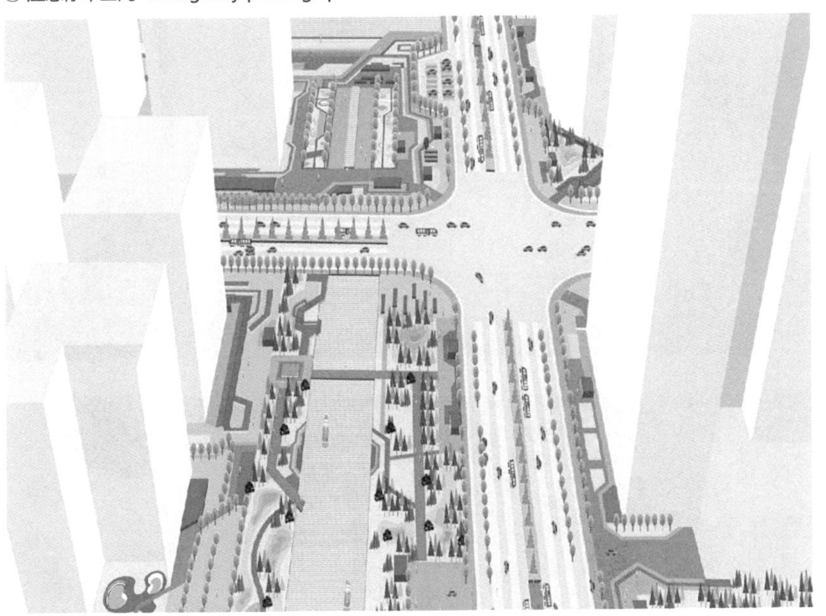

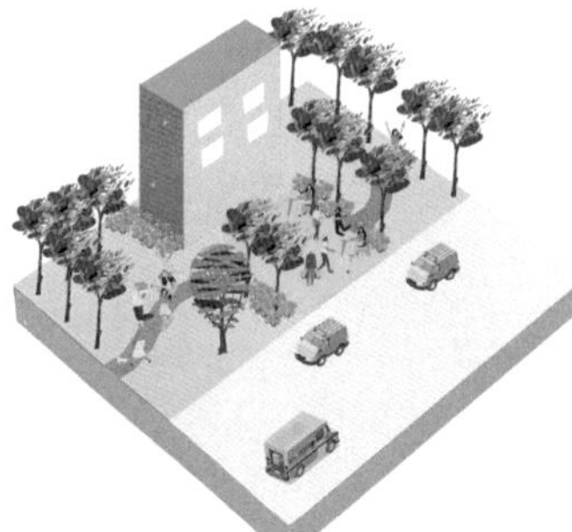

商业活动空间
commercial activity space

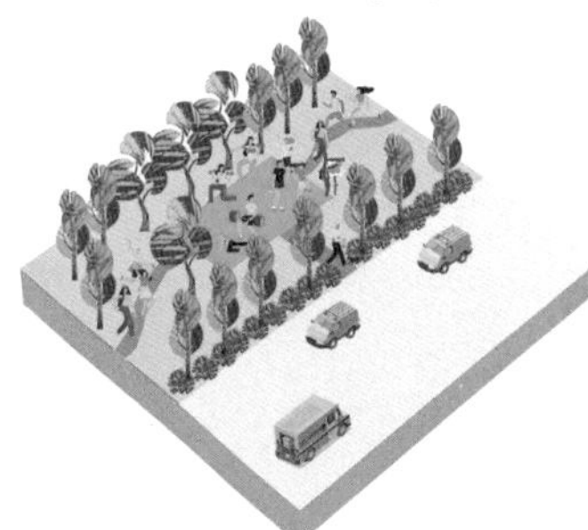

休闲运动空间
leisure and sports space

根据周围建筑的不同性质，营造不同类型的公共活动空间。公交车和公共汽车道路与城市慢行通道有机融合，形成良好的城市交通系统。

According to the different nature of the surrounding buildings, different types of public activity space are created. Buses and bus roads are organically integrated with urban slow lane to form a good urban transportation system.

详细设计 Detailed Design

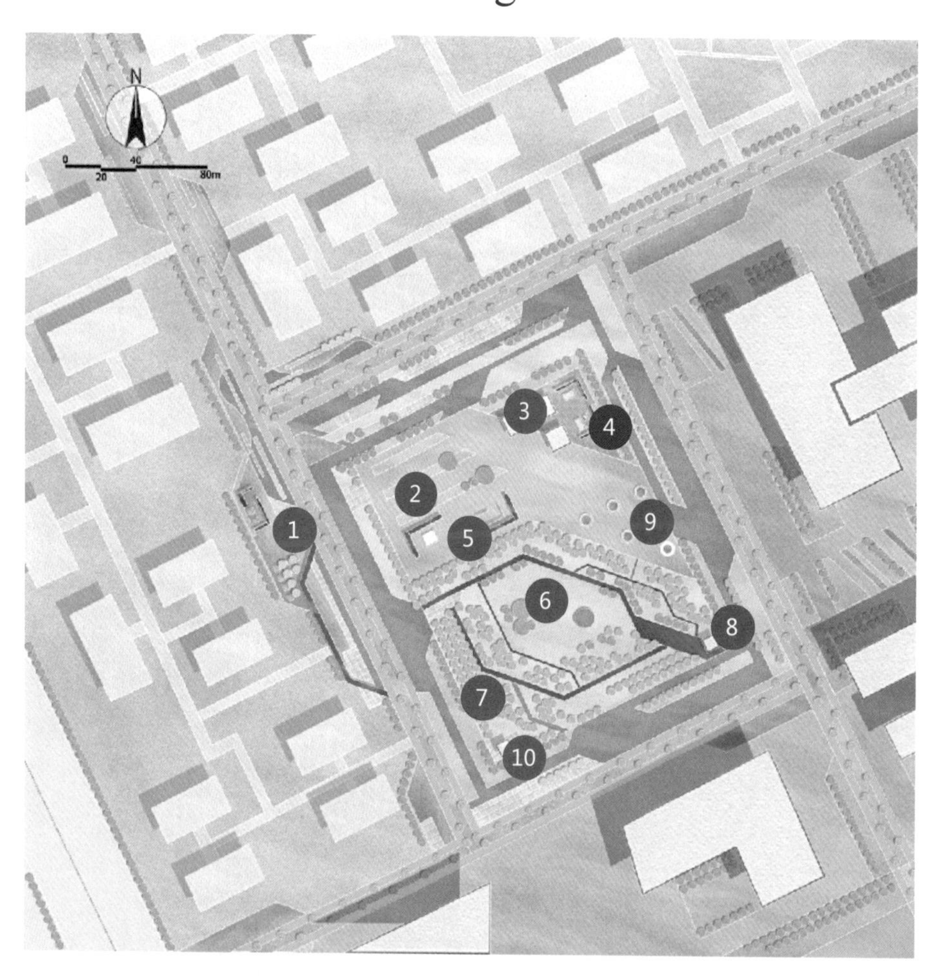

① 街旁休闲场地
street side leisure area

② 临时停车点
temporary parking

③ 汽车专用电梯
car elevator

④ 智慧物流配送点
smart logistics distribution point

⑤ 广场交流空间
square communication space

⑥ 活力大草坪
vibrant lawn

⑦ 林中步道
trail in the forest

⑧ 绿植车行道
green-planted road

⑨ 树池广场
tree square

⑩ 休息小屋
rest cottage

道路分时段管理：高峰期，路旁休憩空间为汽车临时停车点；非高峰期，转换为休憩空间和人们的活动空间。

Roadside open space is used as a temporary parking spot for cars during peak hours and converted to a space for rest and people's activities during off-peak hours.

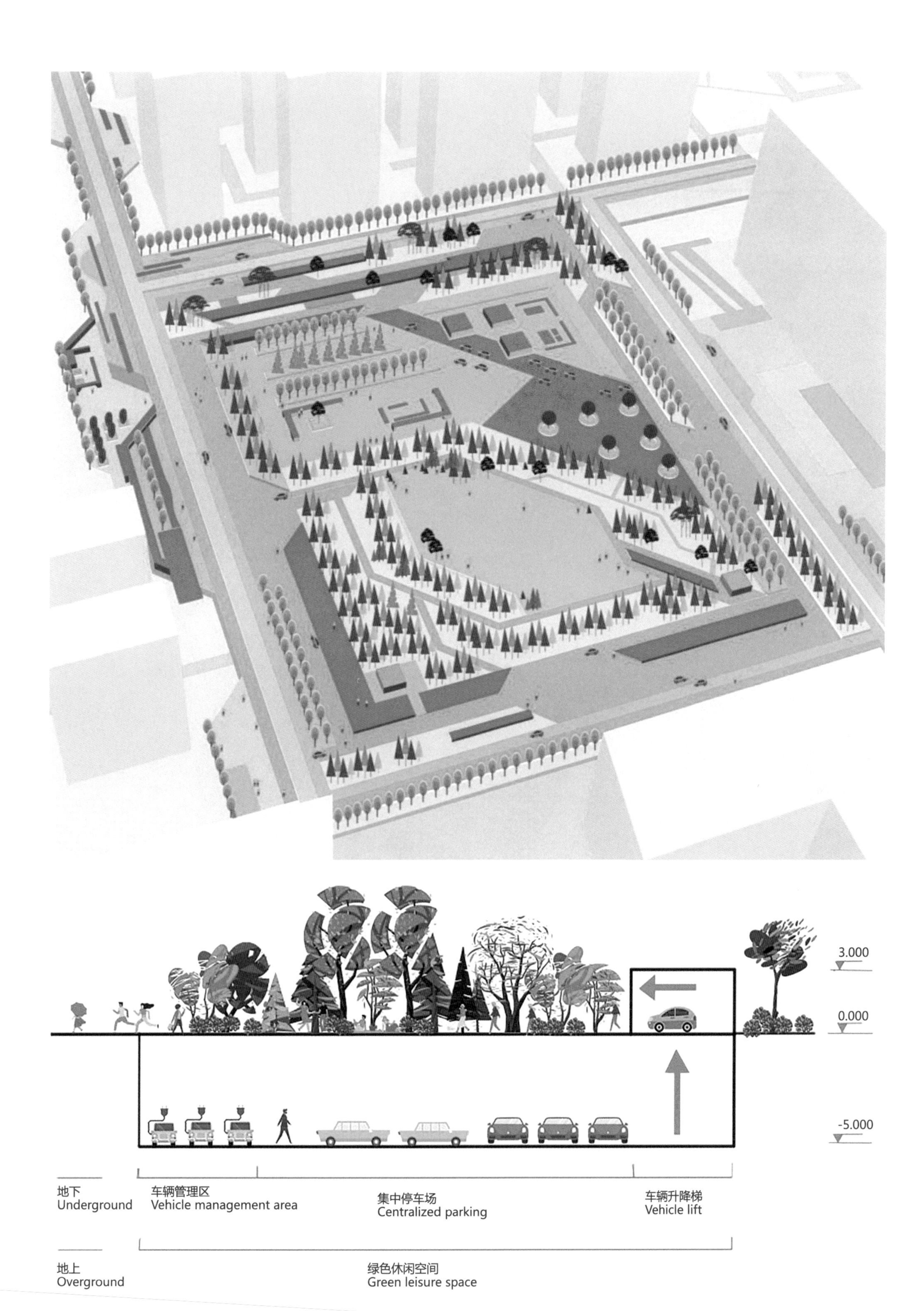
3.000
0.000
-5.000
地下
Underground
车辆管理区
Vehicle management area
集中停车场
Centralized parking
车辆升降梯
Vehicle lift
地上
Overground
绿色休闲空间
Green leisure space

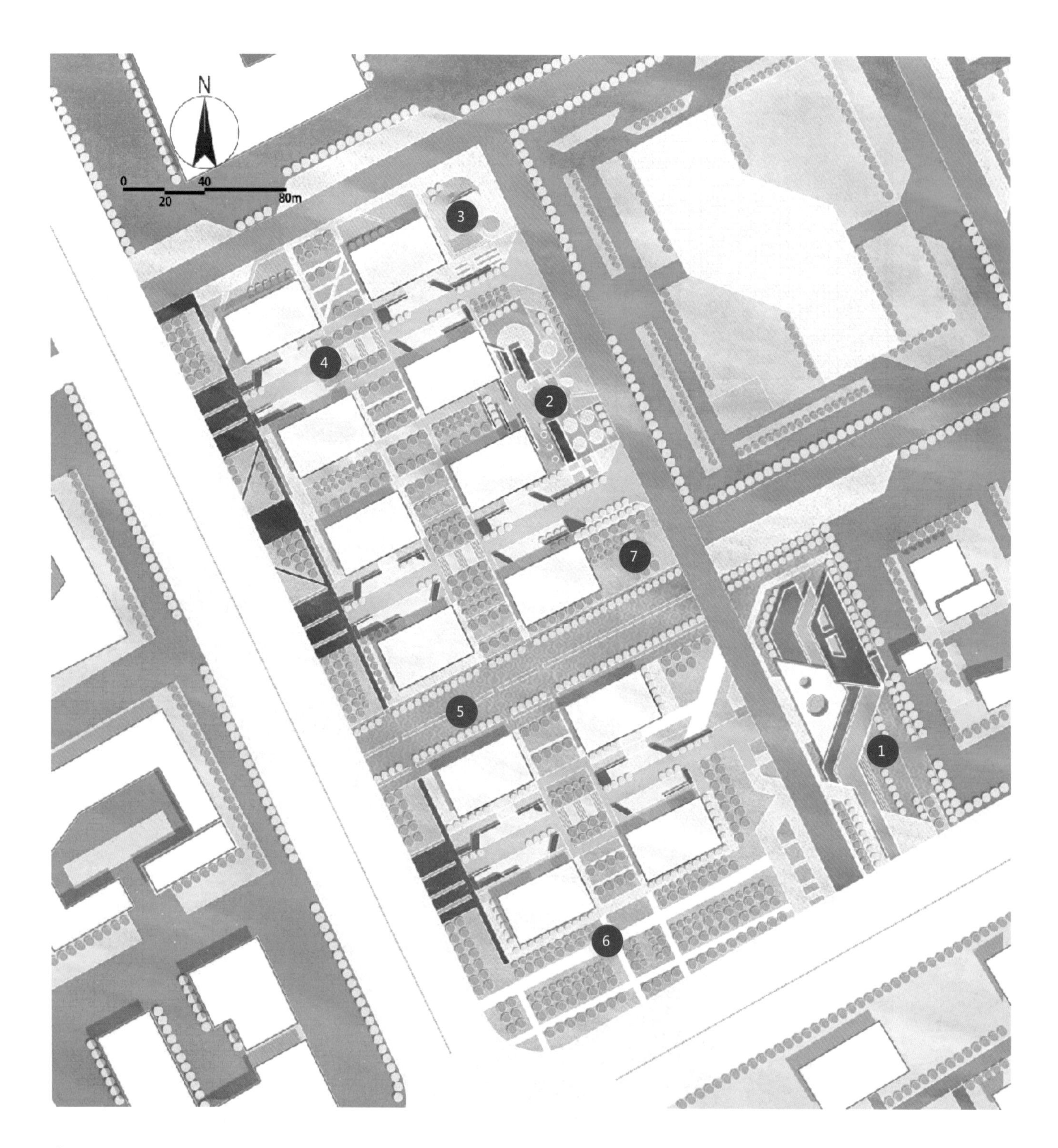

① 户外文化活动室 outdoor cultural activity room
② 社区儿童活动 community children's activities
③ 社区运动场 community playground
④ 停车亭/休憩亭 parking pavilion/lounge
⑤ 社区活动街道 community activity streets
⑥ 社区菜园 community garden
⑦ 社区老年人小游园 community old people's small park

中央绿带与场地中大型商业主轴上的公共空间通过新区大道和谌家矶大道联系起来。无人驾驶汽车可通过穿过主轴的城市支路进入场地内部道路，从而进入商业建筑中。

Through the central green belt of New Area Avenue and Chenjiaji Avenue, it is connected with the public space of the large commercial axis of the site. The driverless vehicle can enter the road inside the site through the urban branch road passing through the main axis, thus entering the commercial buildings.

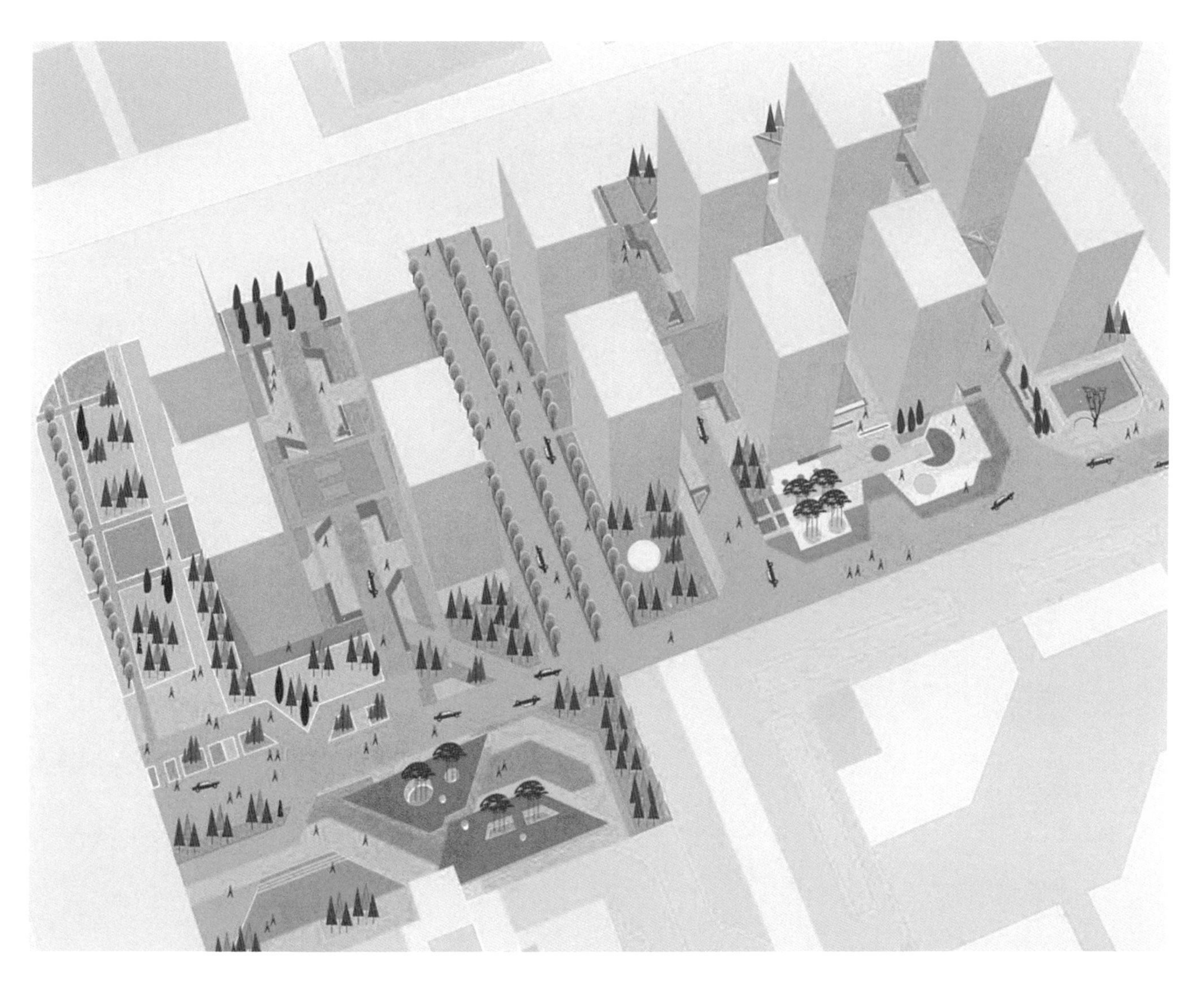

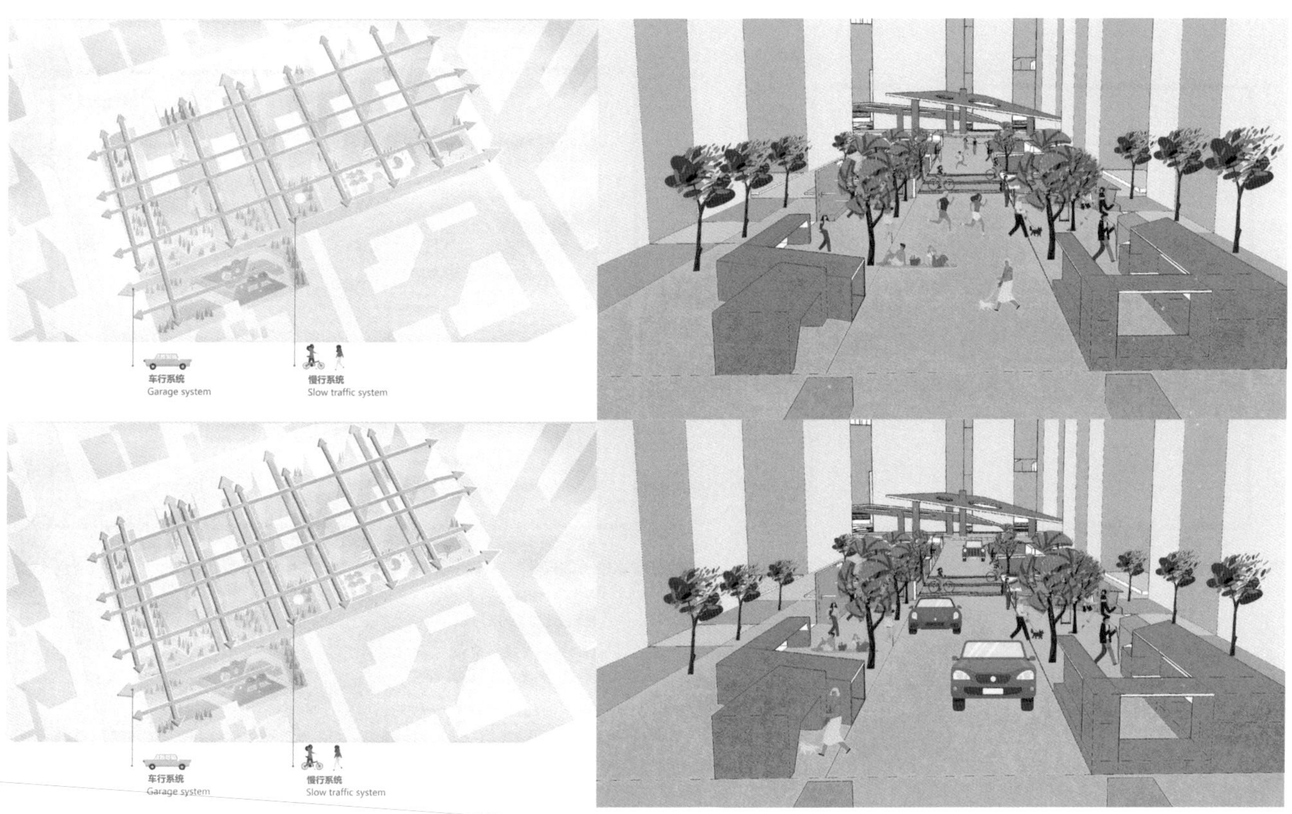
车行系统
Garage system
慢行系统
Slow traffic system
车行系统
Garage system
慢行系统
Slow traffic system

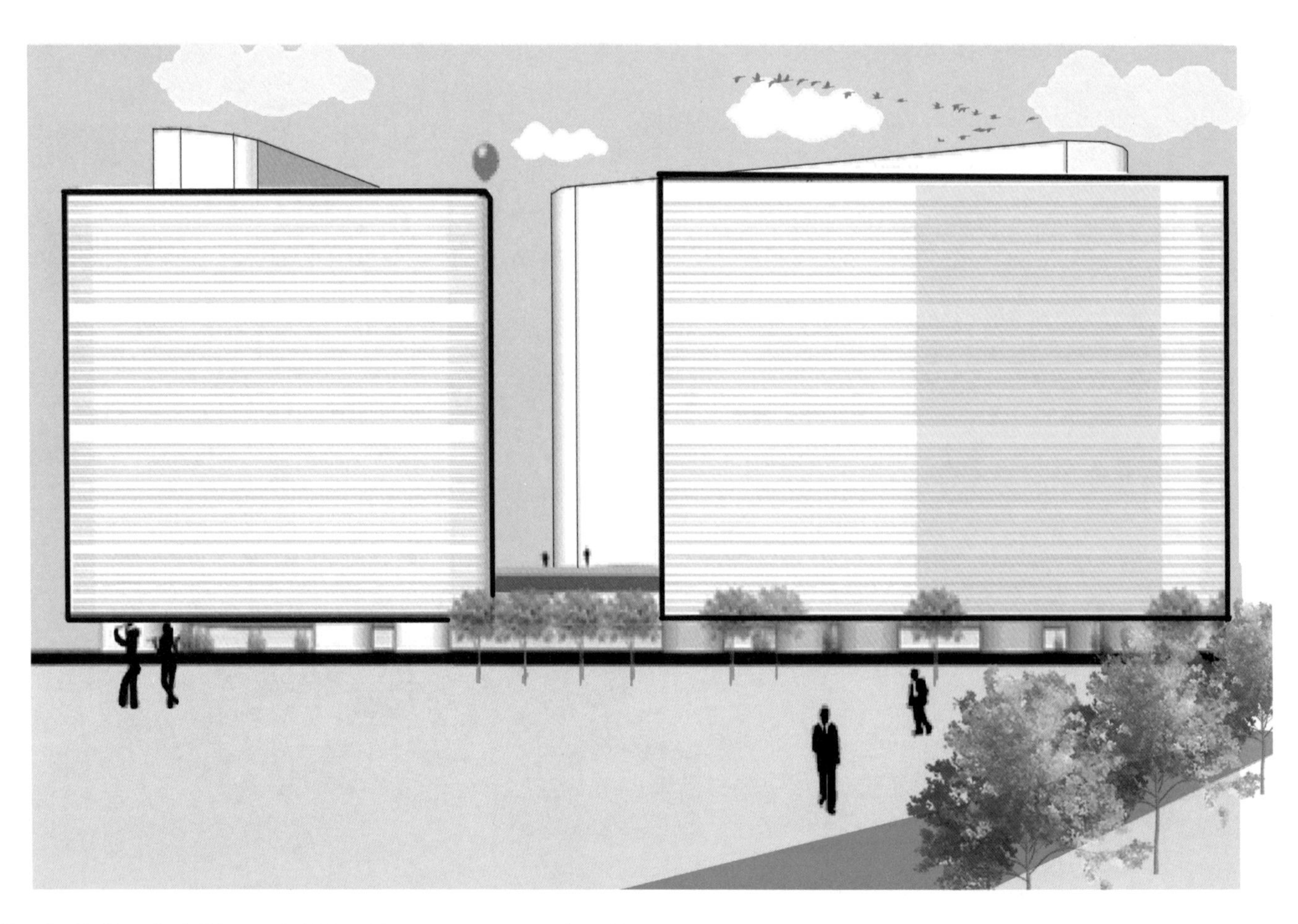

绿色半岛　活力新城

Green Peninsula　Dynamic New Town

陈娜
CHEN Na

江莎
JIANG Sha

刘彦辰
LIU Yanchen

任雨菲
REN Yufei

苏晓丽
SU Xiaoli

杨超
YANG Chao

周媛
ZHOU Yuan

长江新城作为武汉东部门户，长江与府河相交于此，拥有良好的区位优势，交通便利，场地内水体丰富，存在大量的潜在绿色空间和大面积可利用的土地；但它也面临着交通不成体系、绿色系统尚未构建、水质受污染严重、缺乏核心竞争力等问题。构建“一核三轴”“一环五带”“四廊多节点”的结构，打造高端生态居住区、生态活力滨水休闲区、商务休闲核心区、多元体验区、科研文娱休闲区和滨江商务区，以形成健康半岛、活力新城。

Yangtze River New City is located at the intersection of the Yangtze River and the Fu River. As the eastern gateway of Wuhan, it has a good location advantage, convenient transportation, abundant water resources, a large amount of potential green space and a large area of usable land. However, there are also problems such as fragmented transportation, a green system that has not yet been established, serious pollution of water quality, and lack of core competitiveness. Our design construct a structure of “one core, three axes”, “one ring, five belts” and “four corridors and multiple nodes” to create high-end ecological residential areas, ecologically vigorous waterfront leisure areas, core business and leisure areas, multiple experience areas, scientific research, entertainment and leisure areas, and riverside business district to form a healthy peninsula and a vibrant new town.

16
21
12
1
3
2
4
14
6
10
5
7
15
9
8
13

① 中央公园 central park
② BMW中心枢纽 BMW center hub
③ 长江大厦 Yangtze River mansion
④ 购物商街 shopping mall
⑤ 城市T台 city T stage
⑥ 商业景观连廊
commercial landscape corridor
⑦ 企业总部大楼
enterprise headquarter building
⑧ 金融中心 financial center
⑨ 生态体育街 ecological sports street
⑩ SOHO公寓 SOHO apartment
⑪ 产品发布中心 product release center
⑫ 综合邻里中心
integrated neighborhood center
⑬ 军工博物馆
military industry museum
⑭ 朱家河公园 Zhujiahe park
⑮ 科技孵化中心
technology incubation center
⑯ 脉动剧场 pulsating theater
⑰ 生态湿地 ecological wetland
⑱ 活力消落带 energy dissipation belt
⑲ 零碳博物馆 zero carbon museum
⑳ 桥下休闲活动区 activity area under bridge
㉑ 综合医院 general hospital

设计分析 Design Analysis

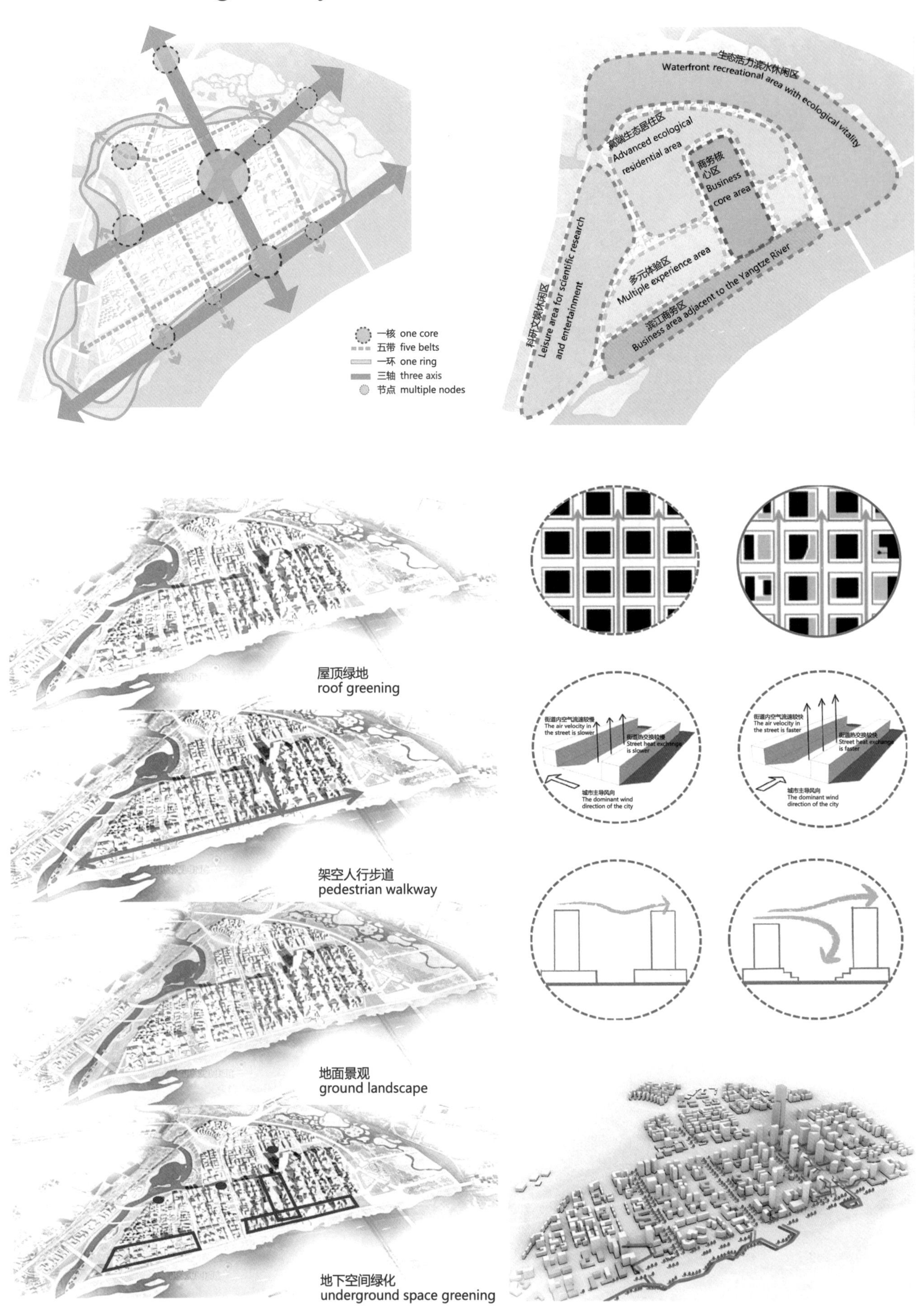

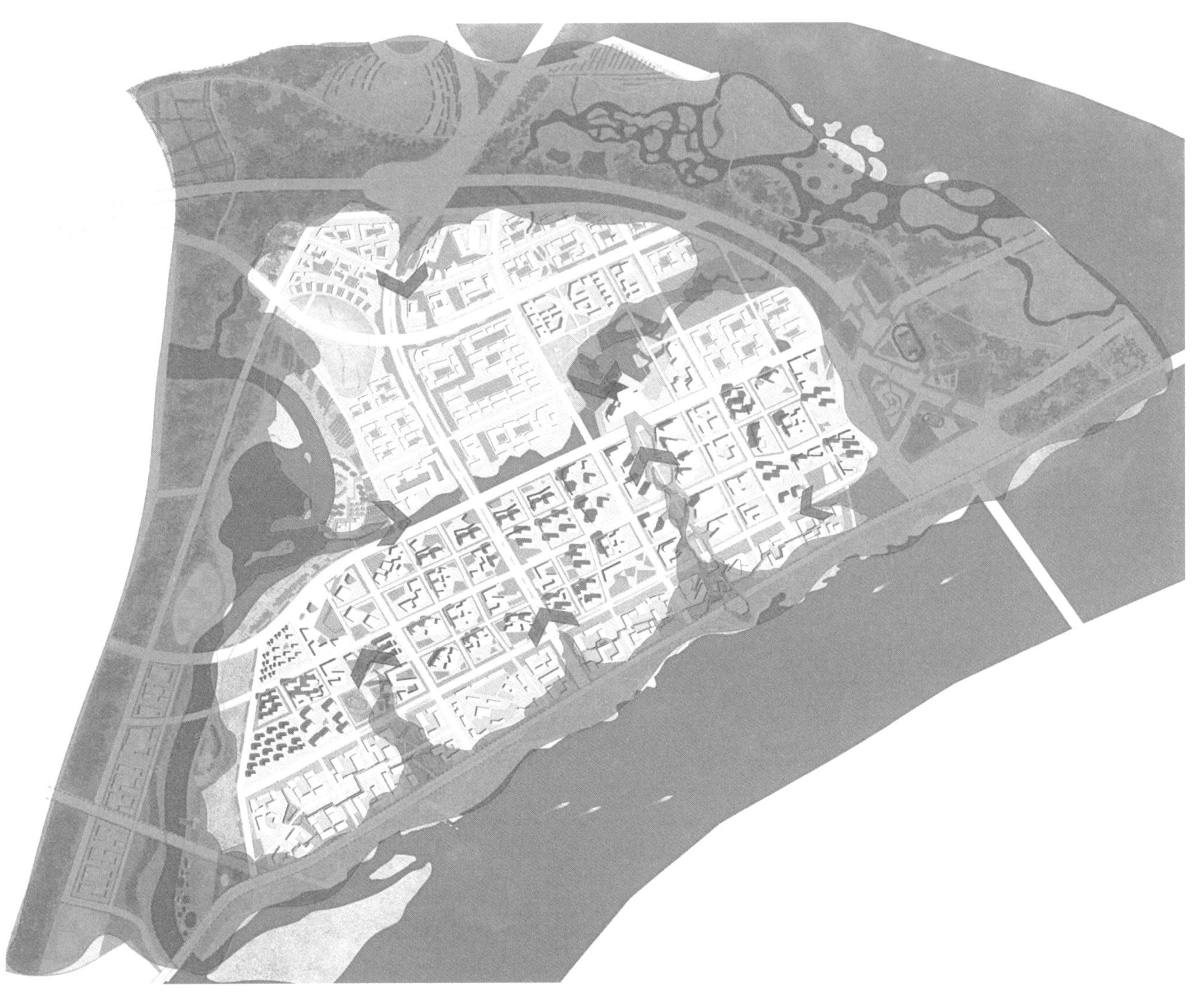

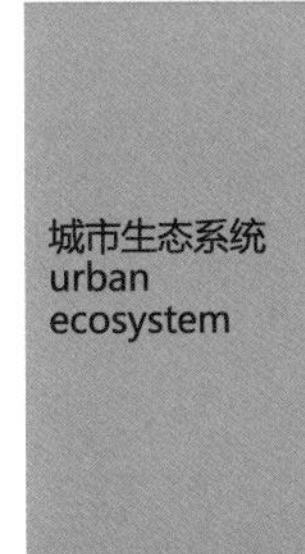

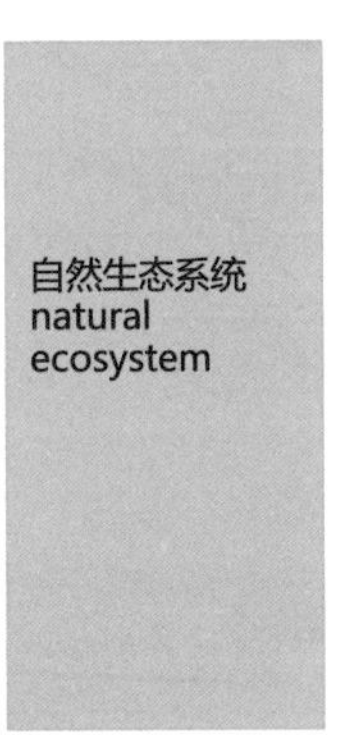

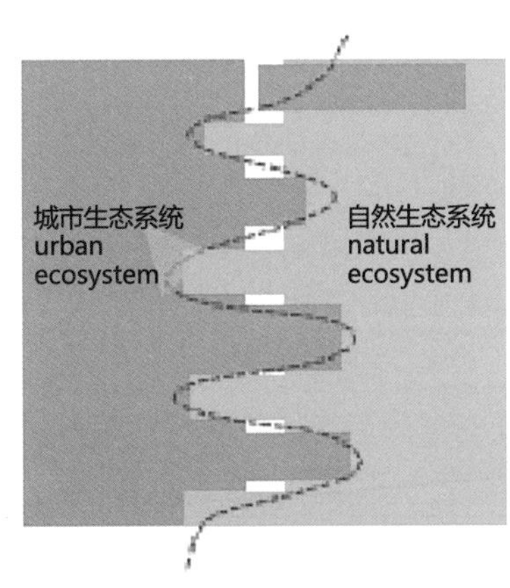

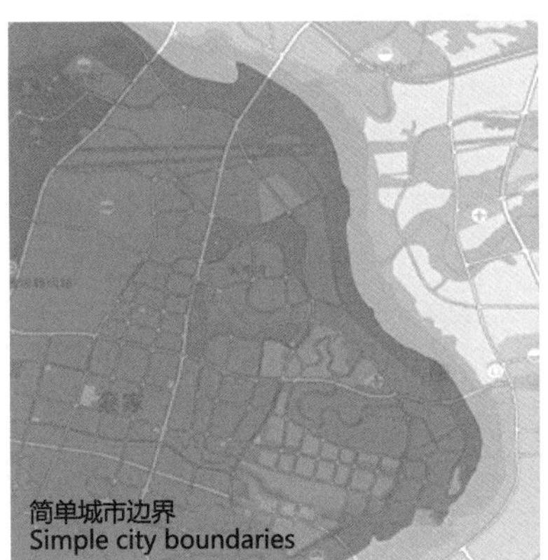

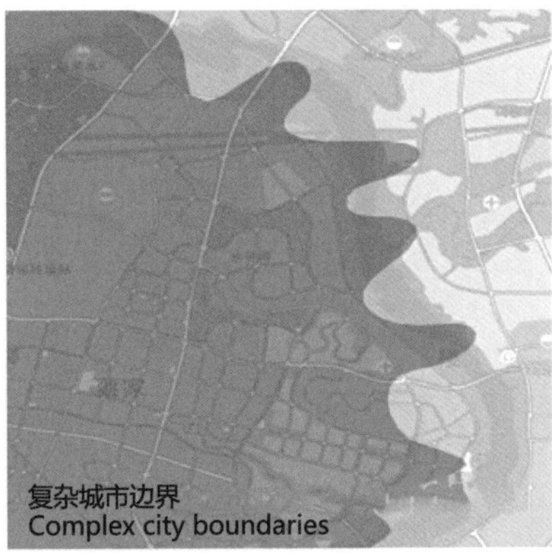

该场地周围的绿色空间向内渗透，以提高边缘的复杂性并缩小城市与自然之间的距离。

河流和绿色空间形成了几个楔形的绿色空间作为载体，以阻止城市之间水平扩展。

绿轴和蓝轴相互连接并渗透，形成动态的景观环。

The green spaces are permeated inward to increase the complexity of the edges and to close the gap between the city and nature.

Several wedge-shaped green spaces are formed by rivers and green spaces as carriers to block the horizontal expansion between cities.

The green axis and the blue axis connect and permeate each other, forming a dynamic ring of landscape.

我们有三种策略来建设绿色交通：将主干道置于地下，使地面道路变窄，并改善地面景观的质量。

We have three strategies to build green transportation. The first is to put the main road underground, narrow the ground road, and improve the quality of the ground landscape.

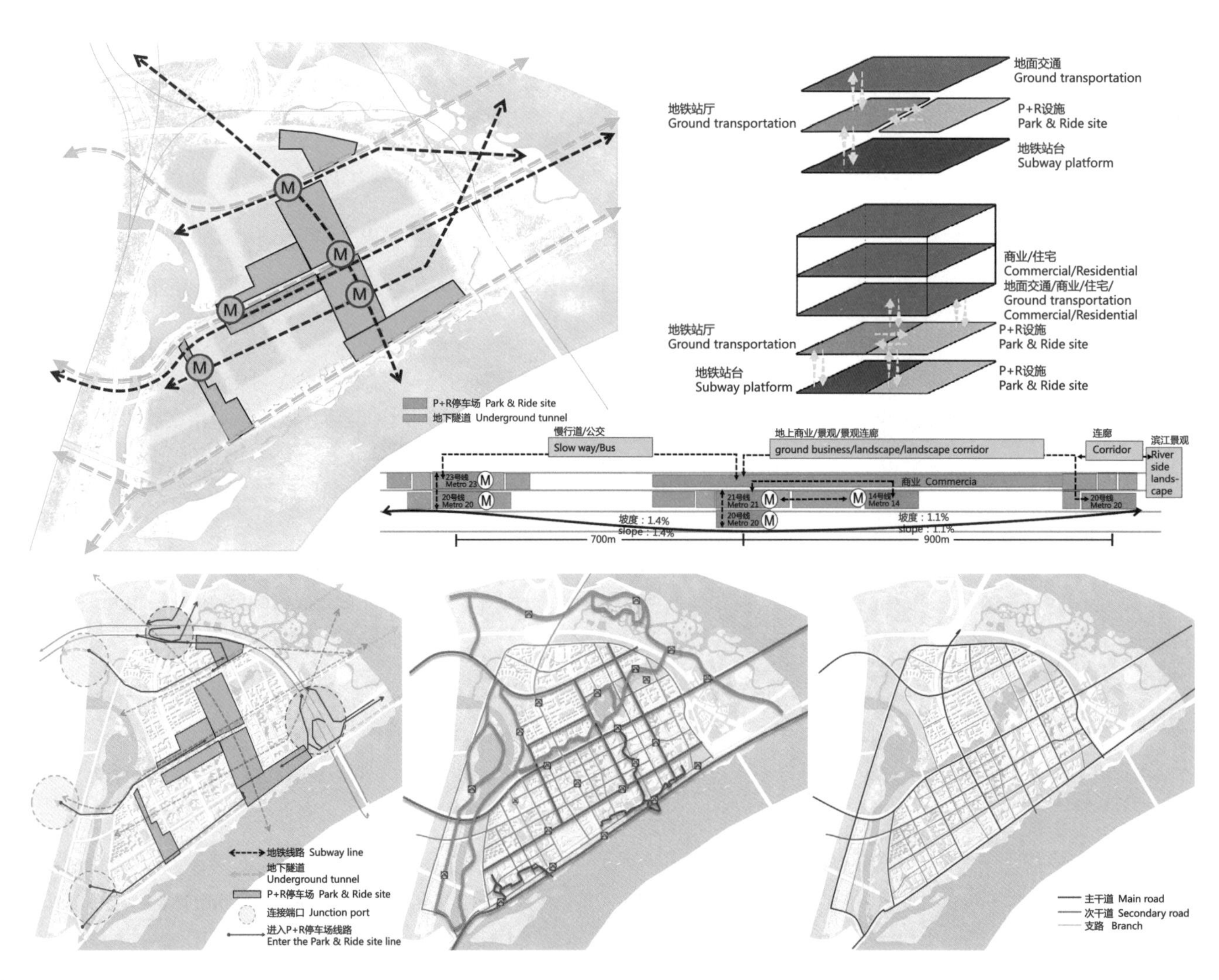

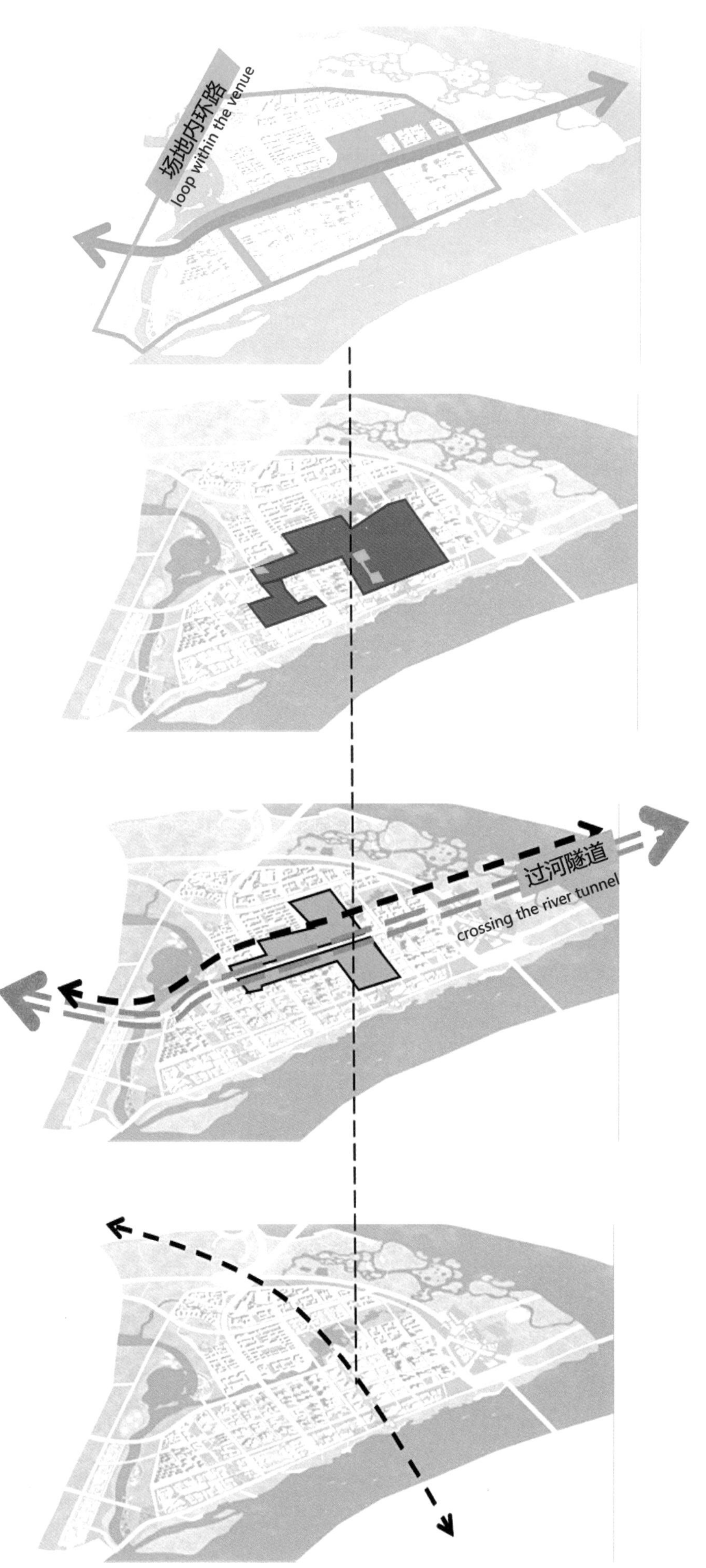

地面：
地面主要用于公共交通。

The ground:
The ground is mainly used by public transportation.

负一层：
负一层分布着地下商业区、地铁车站大厅、下沉广场和绿地。

Basement one:
The basement one is distributed with underground commercial, subway station hall, sinking square and green space.

地下二层：
地铁线位于地下二层的中间，地下隧道与两侧平行。

Basement two:
The subway line is in the middle of the basement two, the underground tunnels are parallel to the two sides.

地下三层：
地铁 20 号线和地下三层的换乘站。

Basement three:
The subway line 20 and subway transfer station are in the basement three.

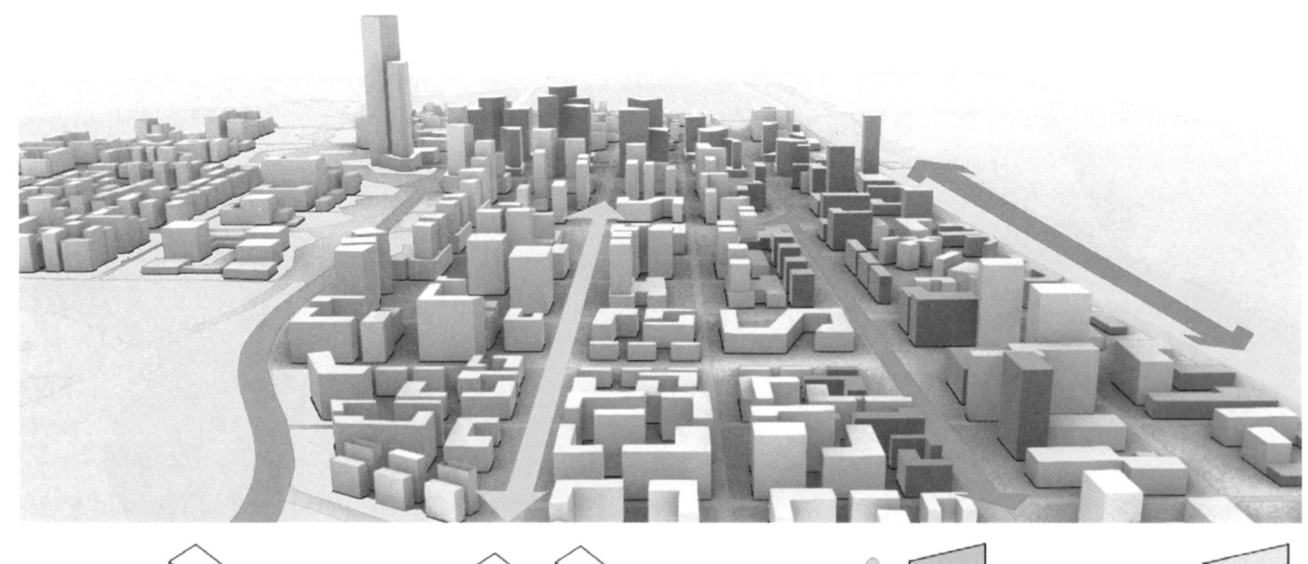

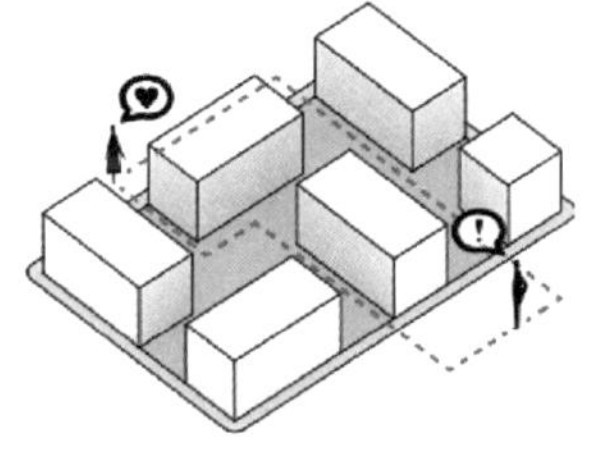

开放街道
open block

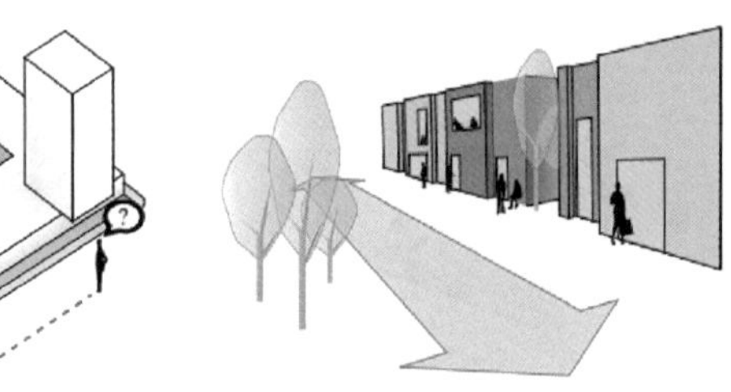

首层可渗透街道界面
permeable street interface on the first floor

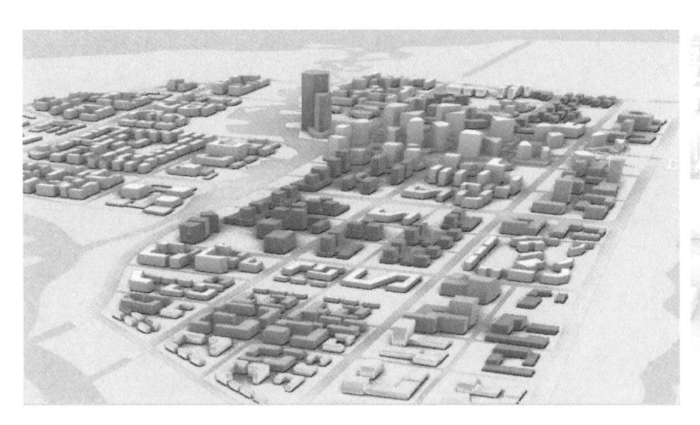

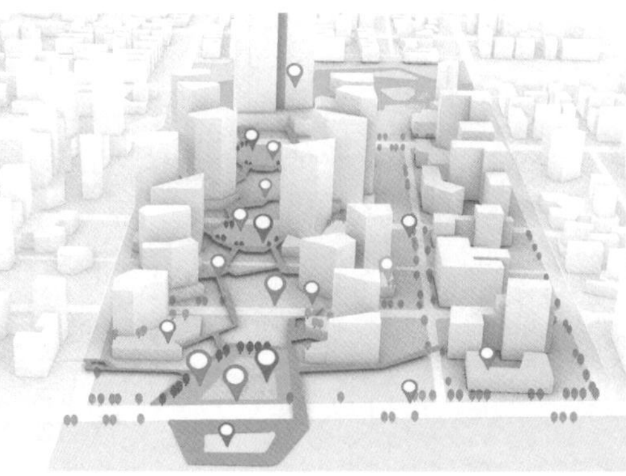

重点区域详细设计 Detailed design of key areas

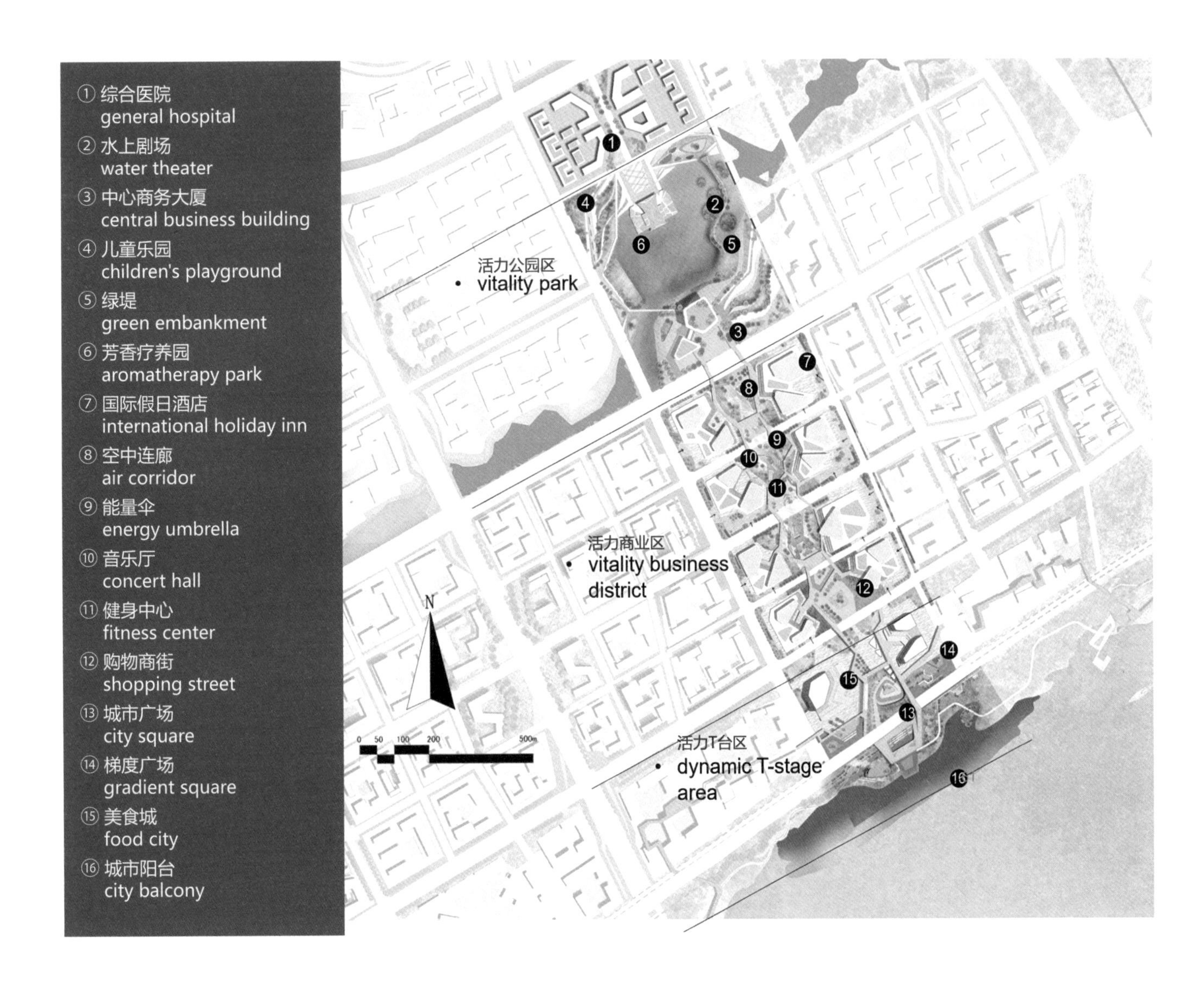

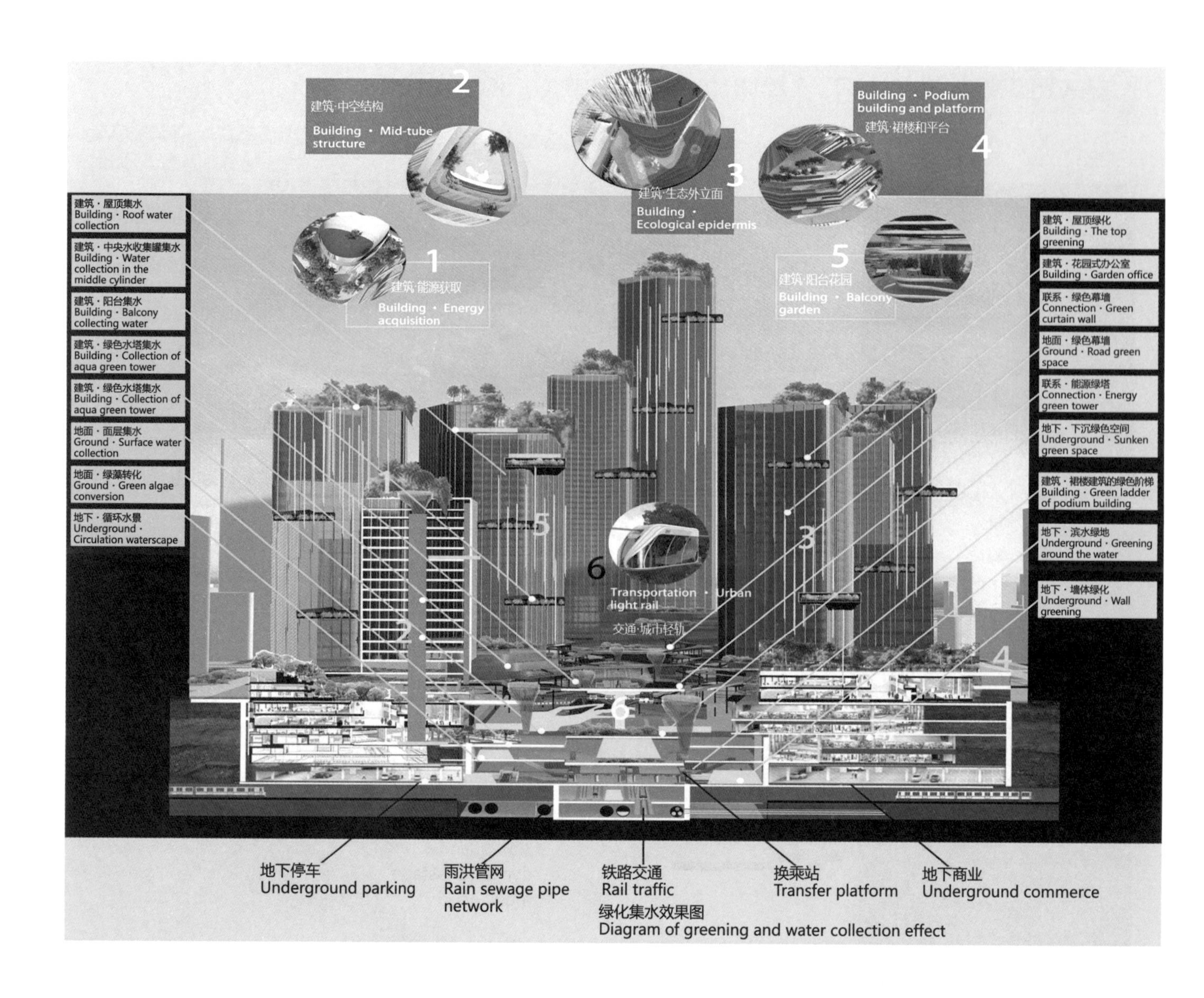

绿化集水效果图
Diagram of greening and water collection effect

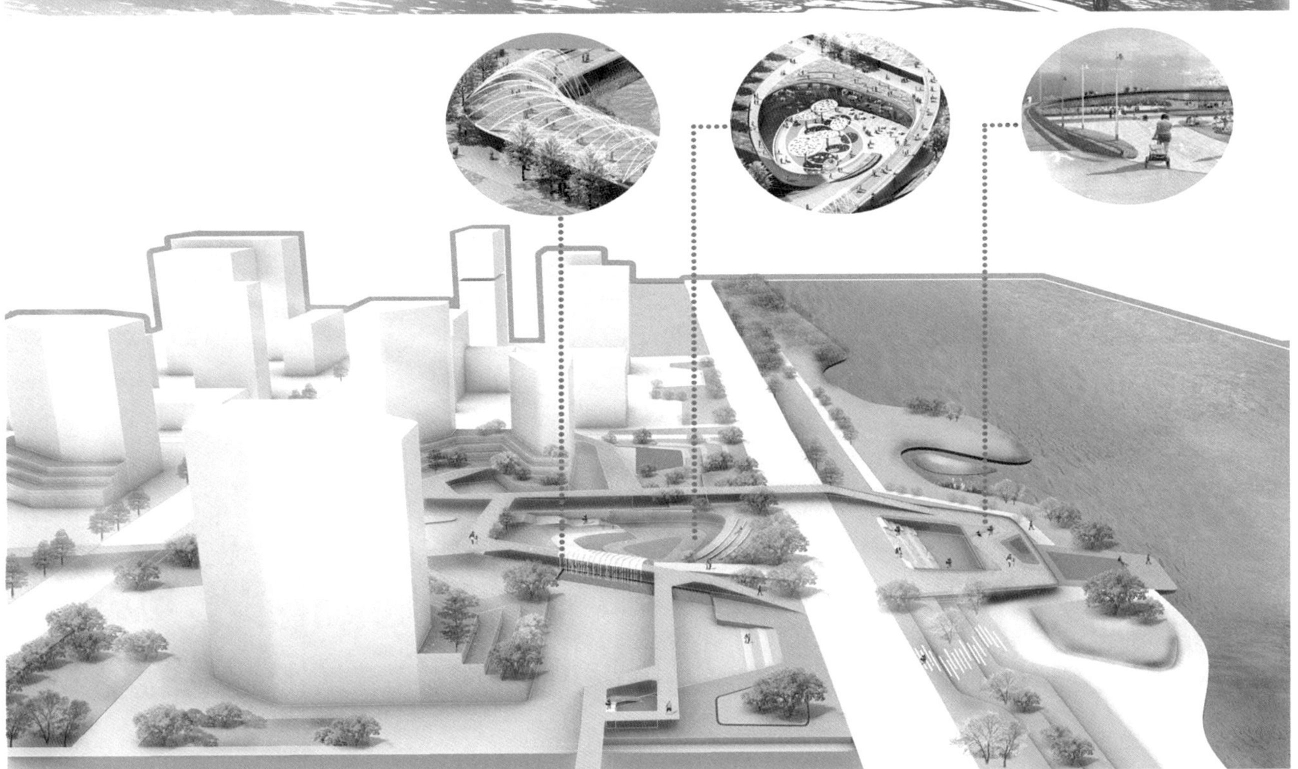

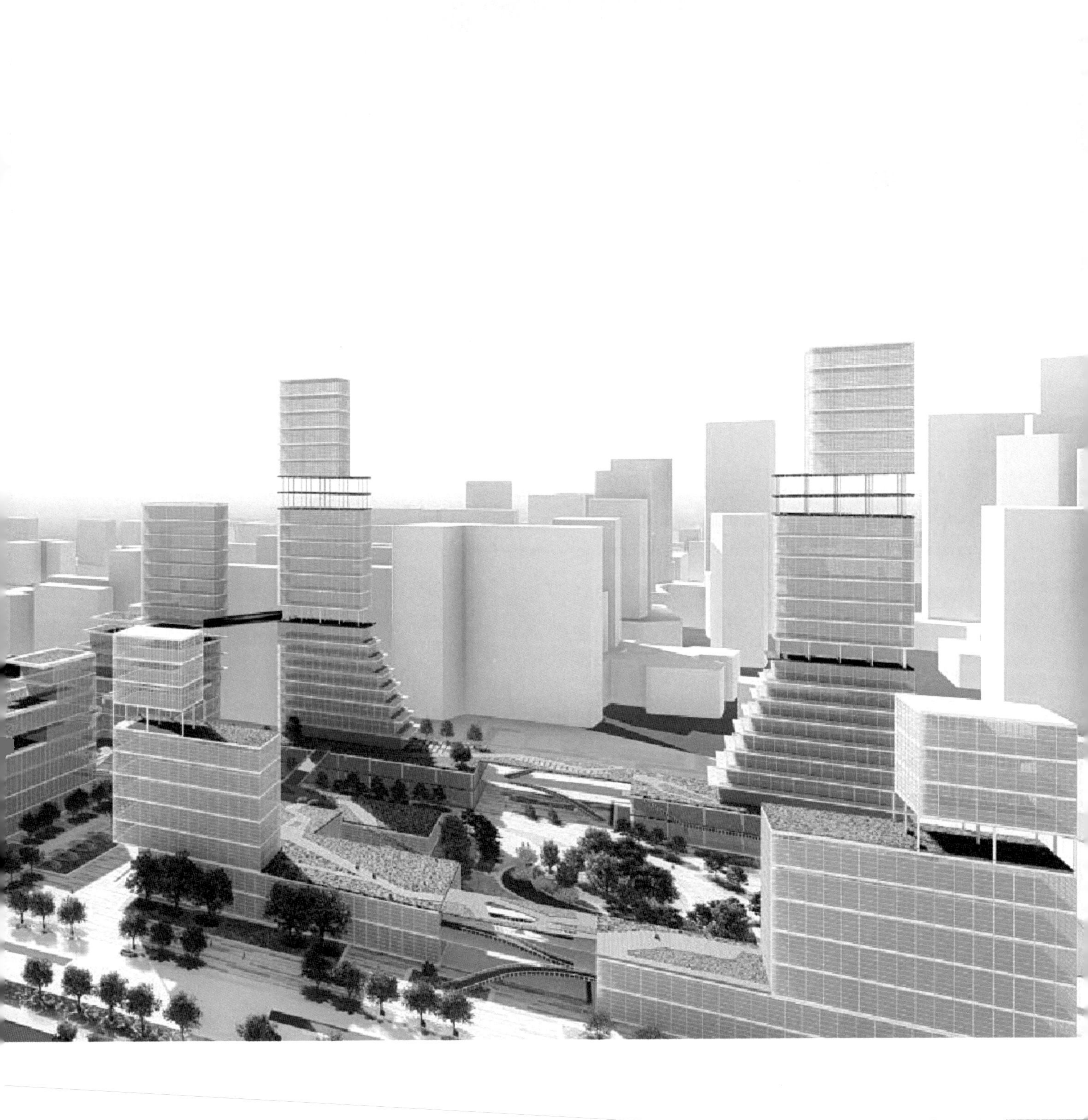

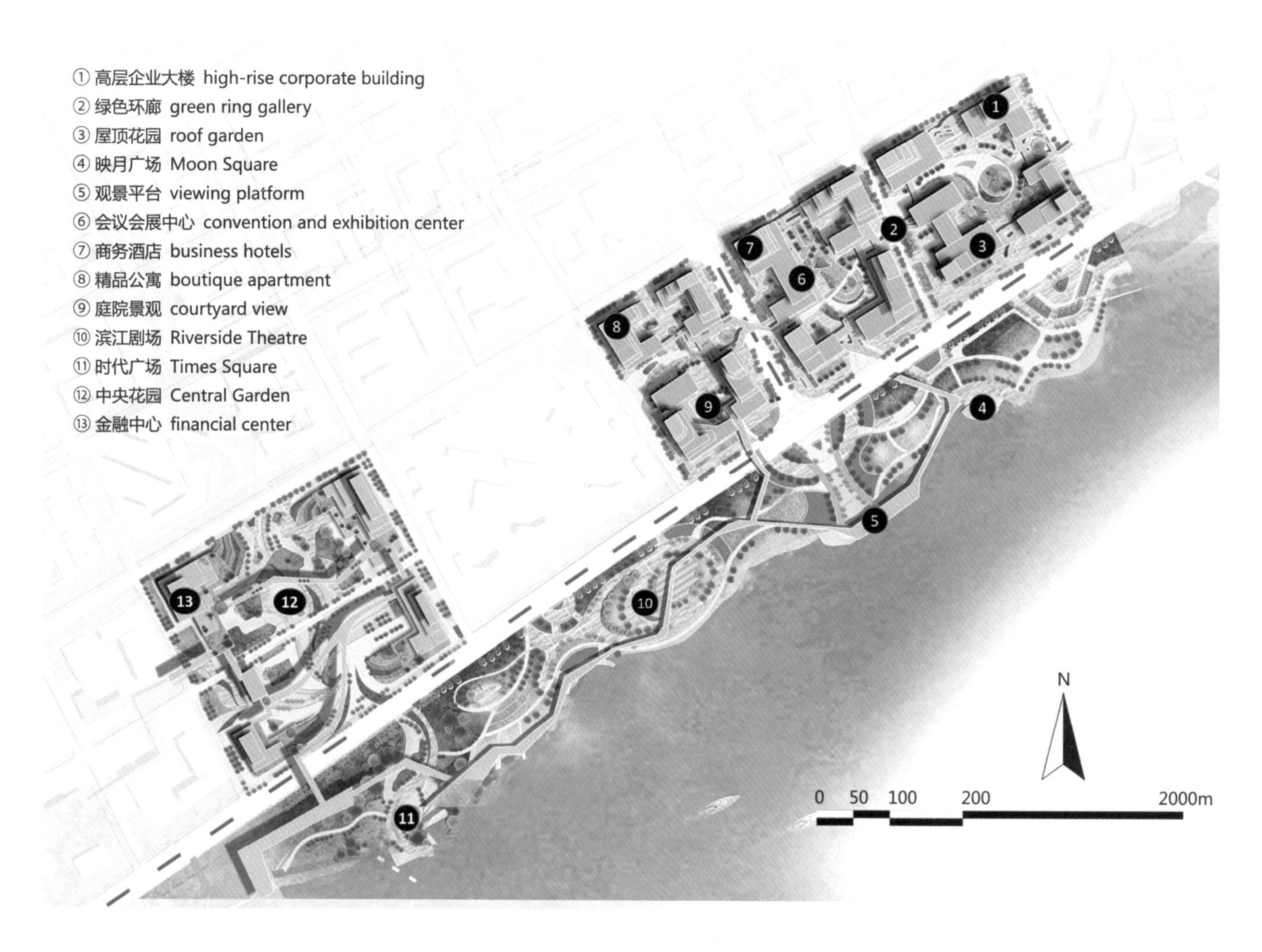

通过土地的混合使用和建筑竖向功能的复合，形成集办公、商业、商务会展、居住于一体的滨江建筑组团。

Through the mixed use of land and the combination of the vertical functions of the building, a riverside architecture group integrating office, commerce, business exhibition and residence is formed.

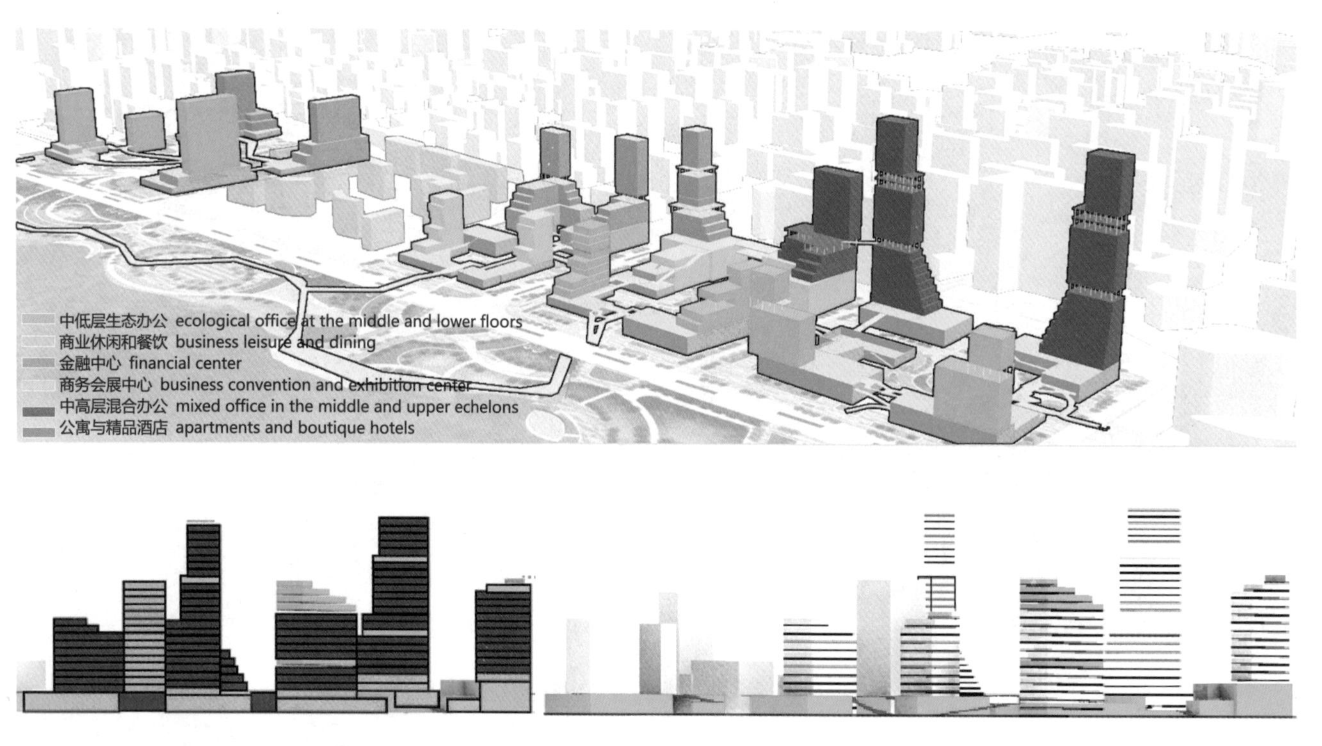

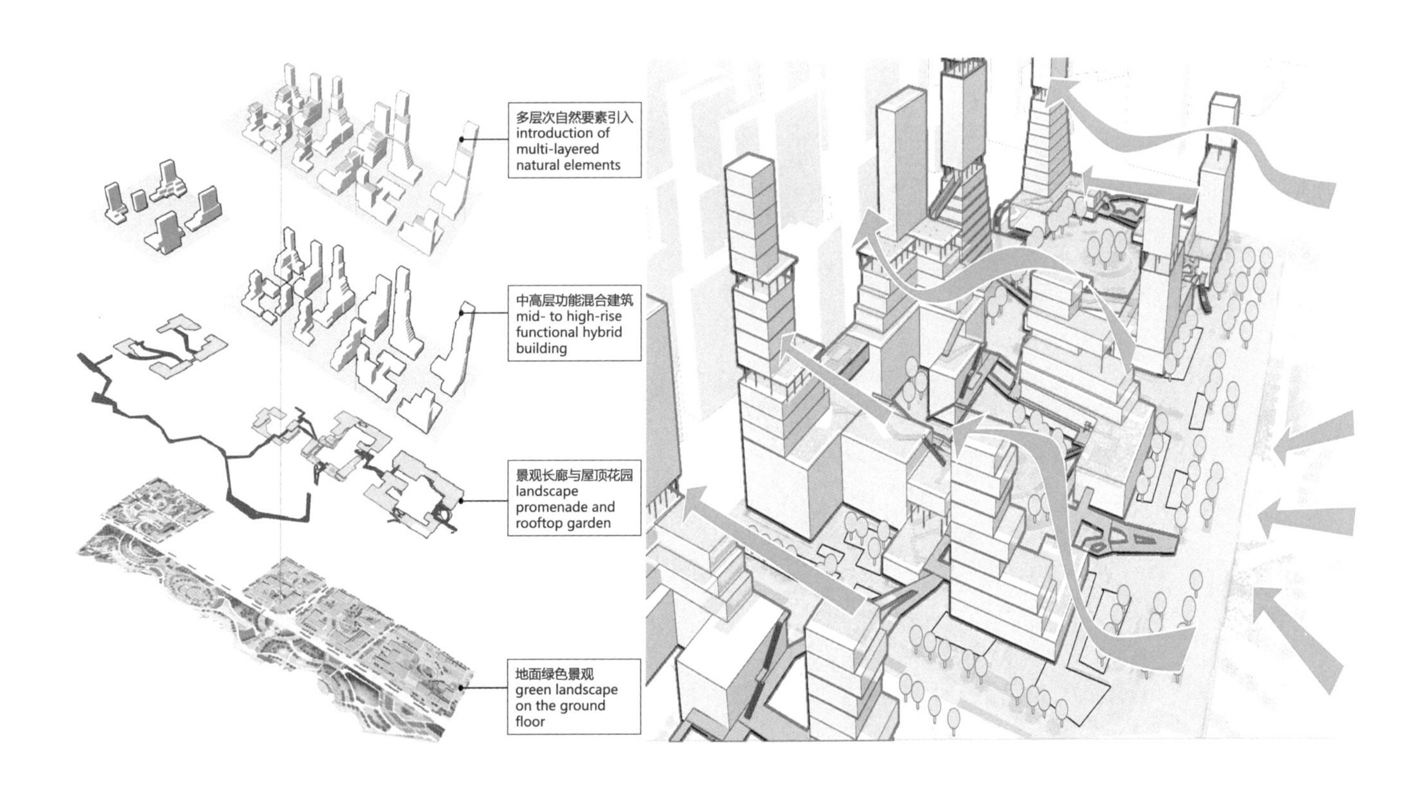
多层次自然要素引入
introduction of multi-layered natural elements
中高层功能混合建筑
mid- to high-rise functional hybrid building
景观长廊与屋顶花园
landscape promenade and rooftop garden
地面绿色景观
green landscape on the ground floor

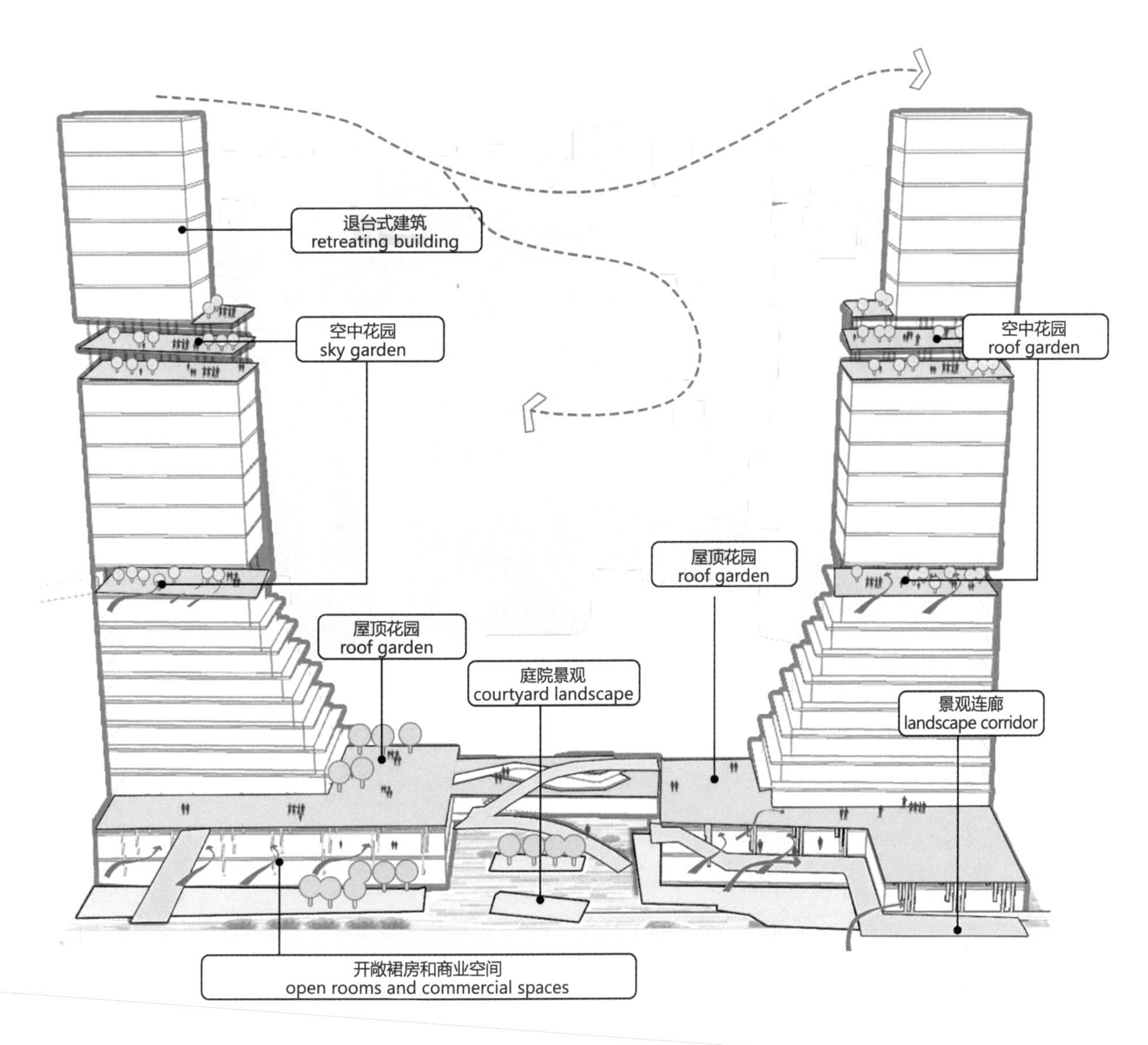
退台式建筑
retreating building
空中花园
sky garden
空中花园
roof garden
屋顶花园
roof garden
屋顶花园
roof garden
庭院景观
courtyard landscape
景观连廊
landscape corridor
开敞裙房和商业空间
open rooms and commercial spaces

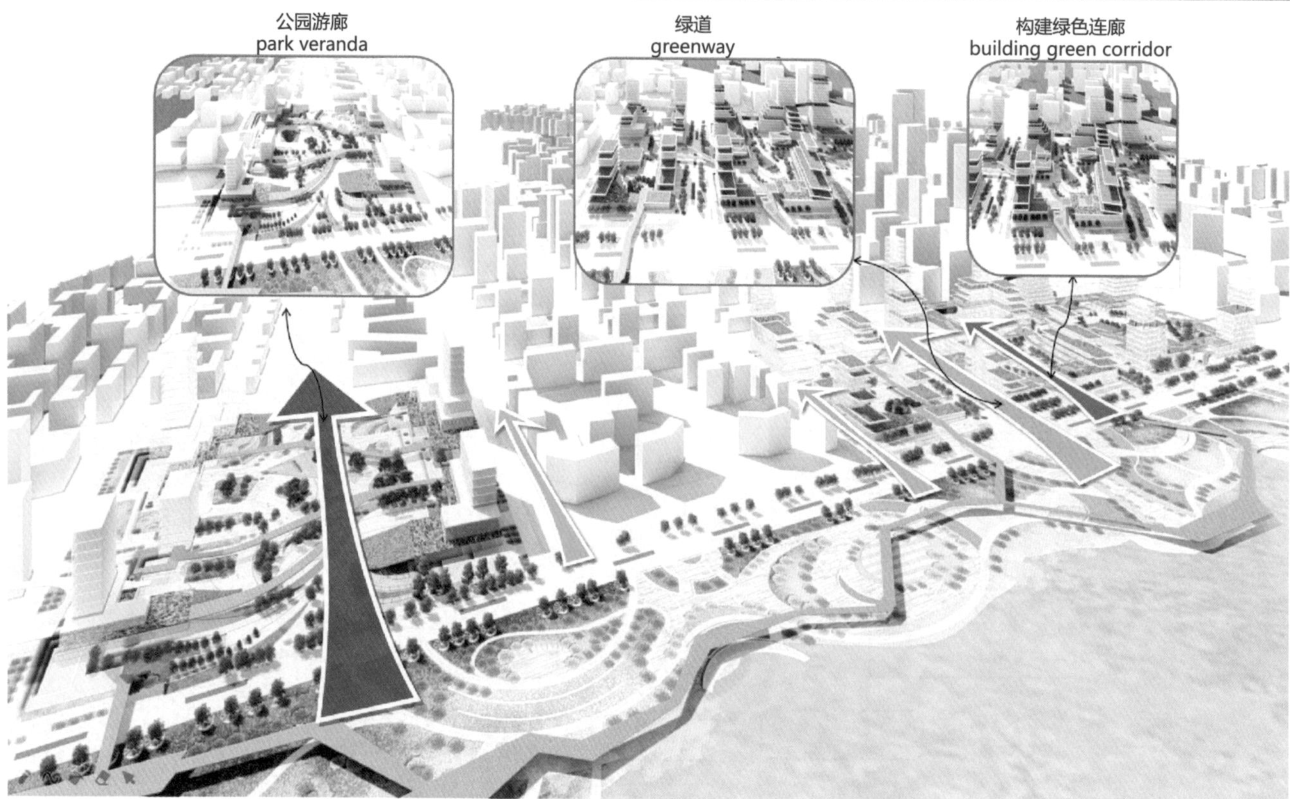

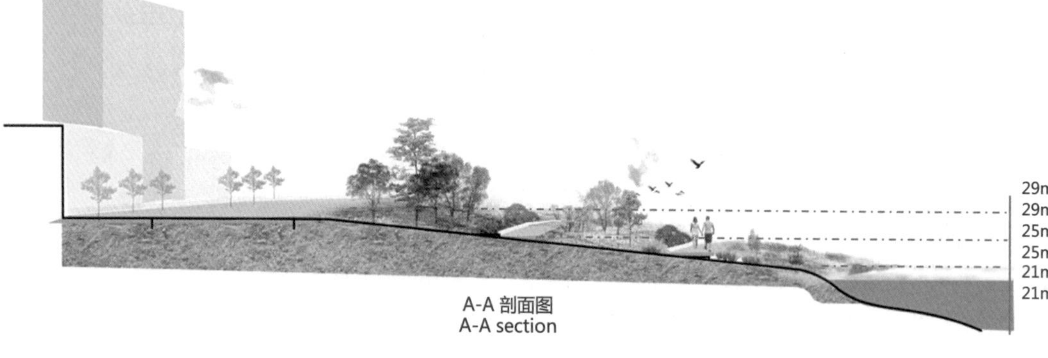

A-A 剖面图
A-A section

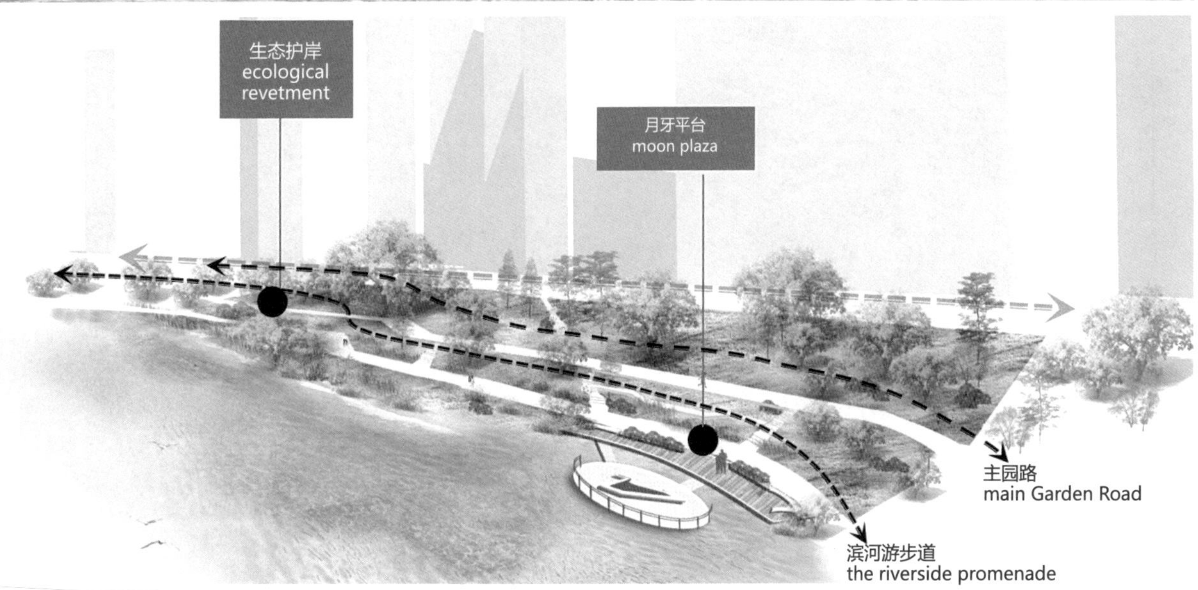
生态护岸
ecological
revetment
月牙平台
moon plaza
主园路
main Garden Road
滨河游步道
the riverside promenade

无界之城

No Boundaries

鲁曦冉
LU Xiran

孙译远
SUN Yiyuan

杨民阁
YANG Minge

李思韬
LI Sitao

雷可欣
LEI Kexin

商晔
SHANG Ye

吴丛冉
WU Congran

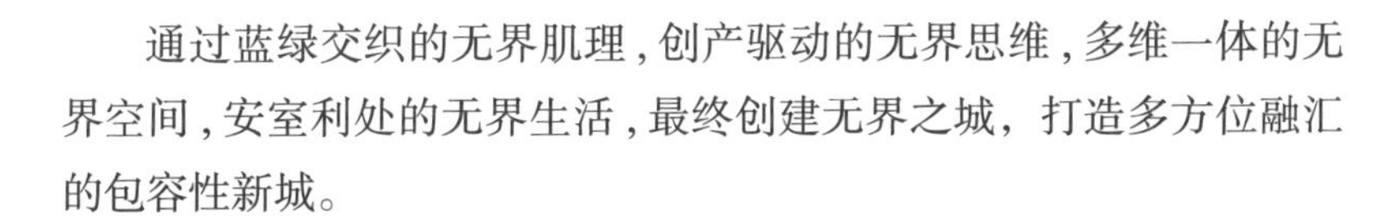

通过蓝绿交织的无界肌理，创产驱动的无界思维，多维一体的无界空间，安室利处的无界生活，最终创建无界之城，打造多方位融汇的包容性新城。

We try to create an unbounded city and a multi-faceted inclusive new city through blue and green intertwined texture, the unbounded creates thinking driven and multi-dimensional unbounded space and life.

鸟瞰图 Aerial View

设计策略 Design Strategy

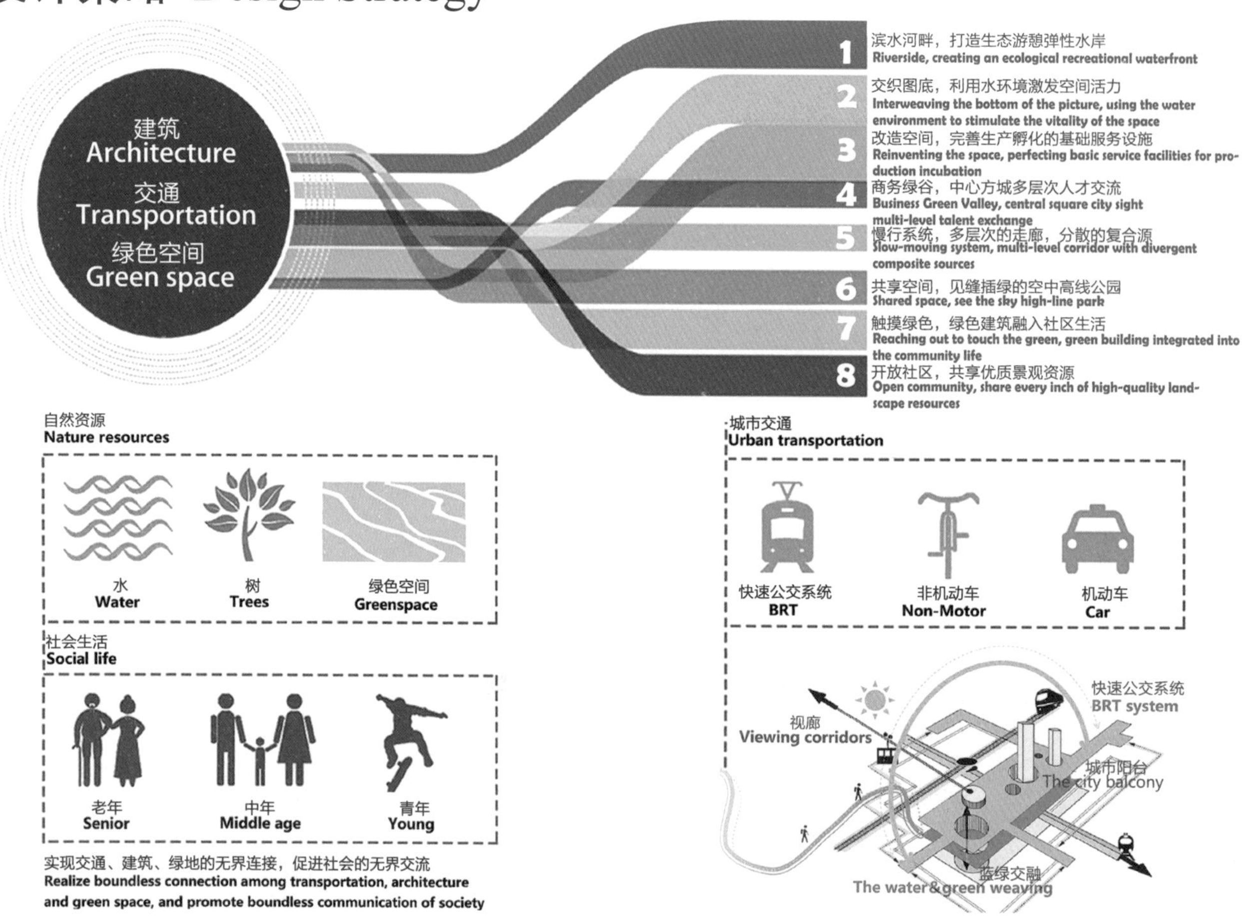

实现交通、建筑、绿地的无界连接，促进社会的无界交流
Realize boundless connection among transportation, architecture and green space, and promote boundless communication of society

营造灵活生态的水岸线
Create a flexible waterfront for ecological recreation

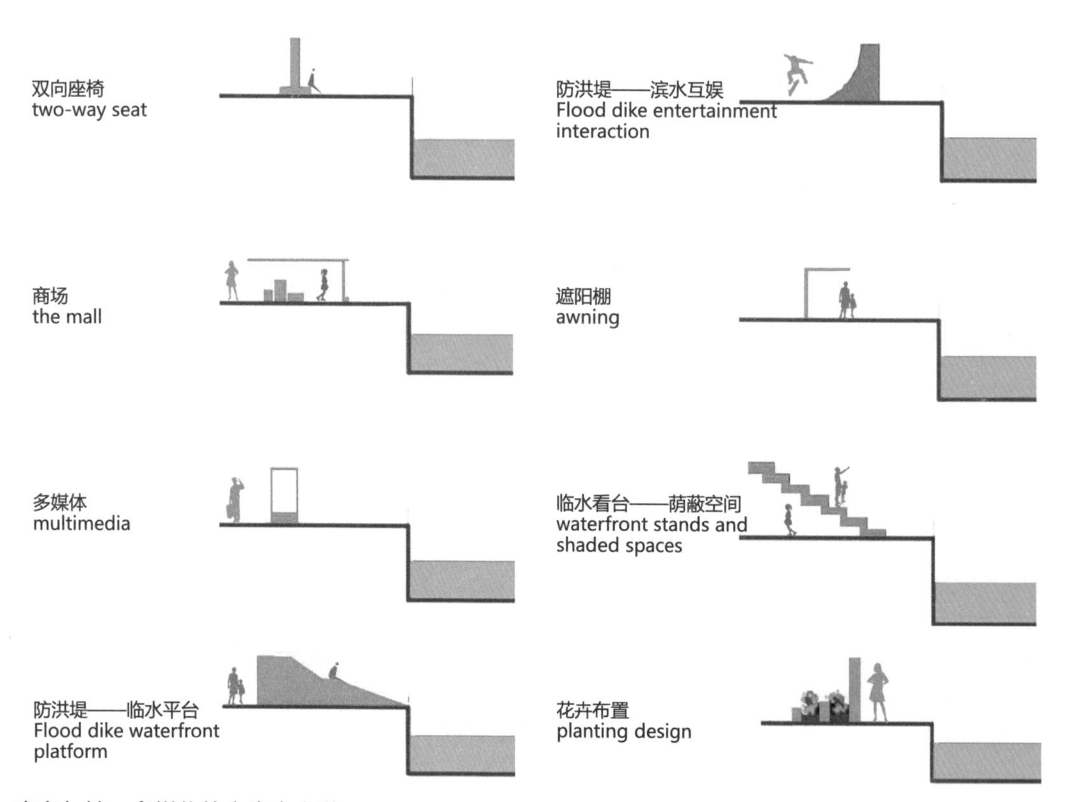

建立安全、高参与性、多样化的水生态廊道
Establish safe, highly participatory and diverse aquatic ecological corridors

“商业绿谷”实现多层联动 "Business Green Valley" realizes multi-level linkage

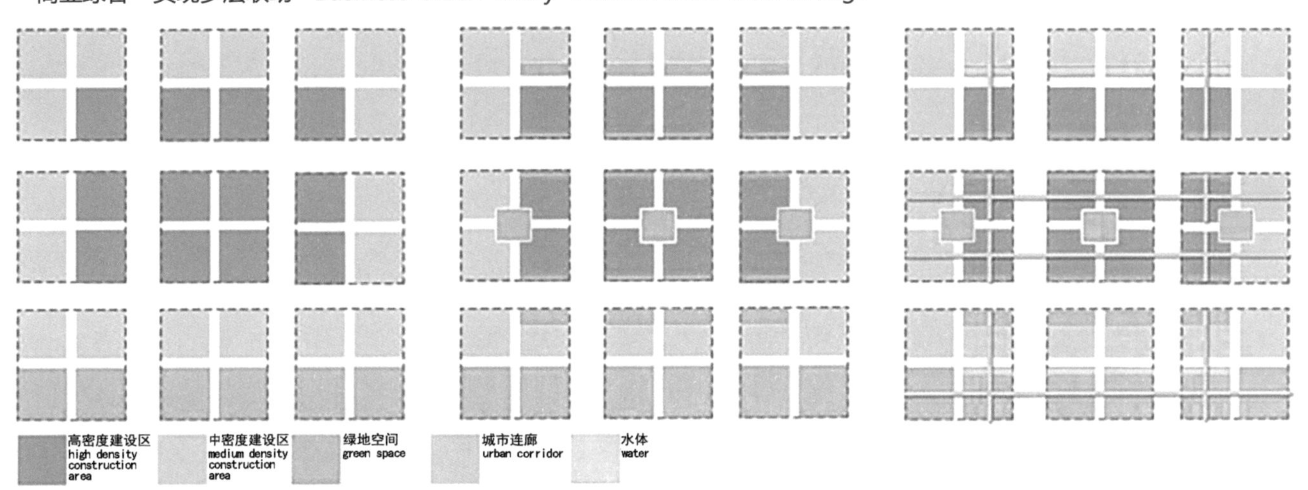

复合立体步行廊道 Composite three-dimensional walking corridor

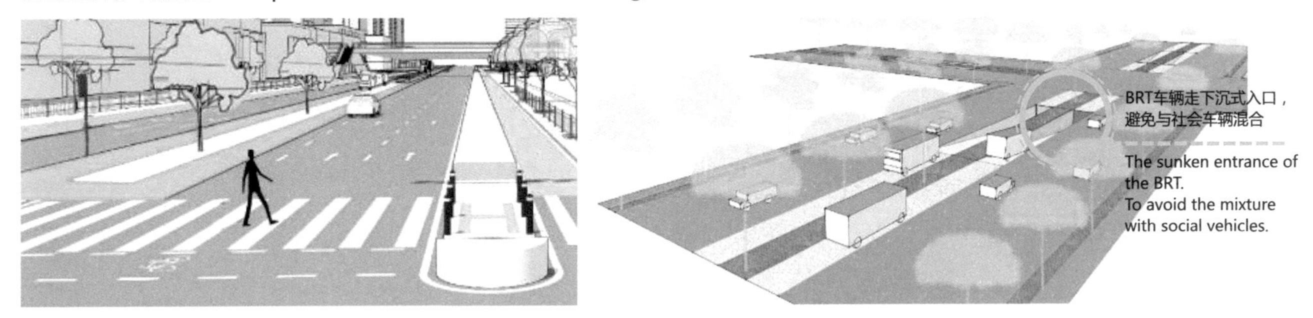

绿色建筑融入日常生活 Green building integrated into daily life

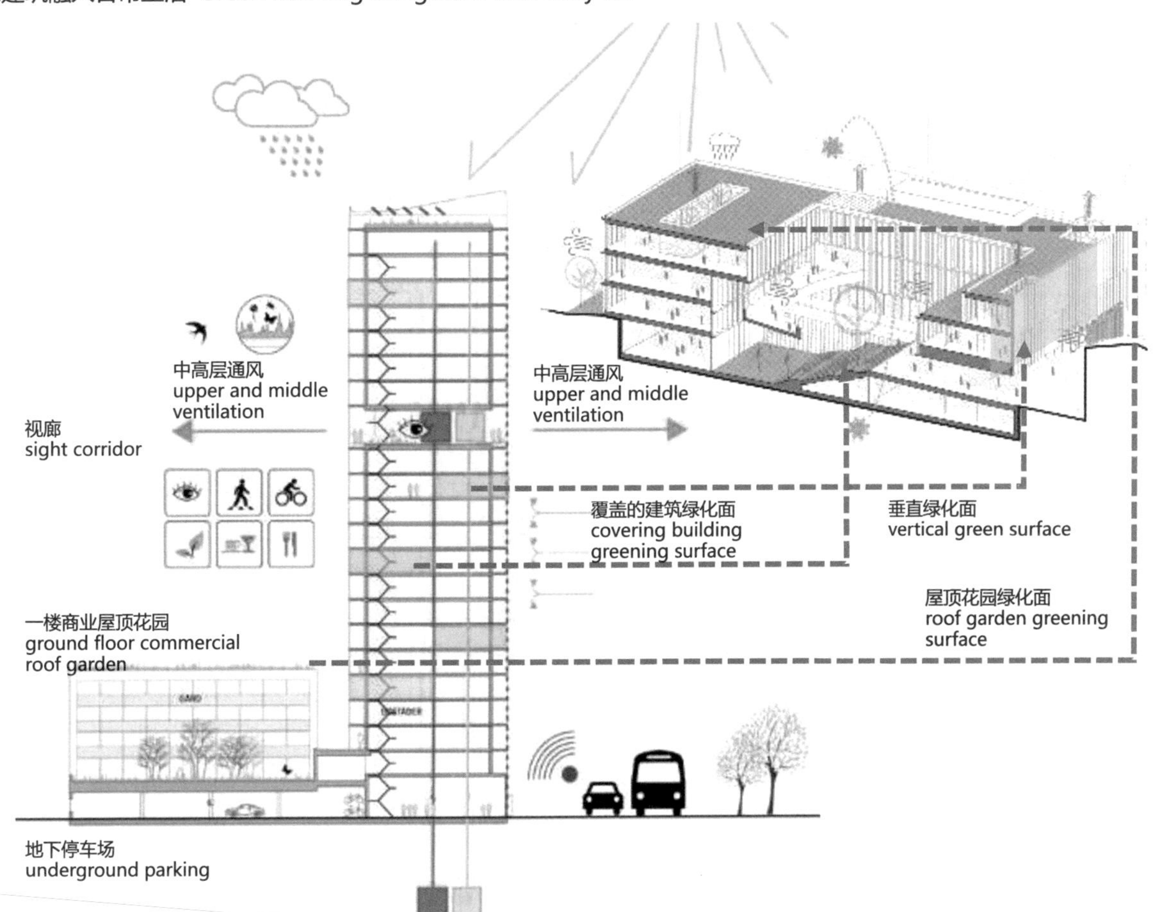

概念推衍 Concept Generation

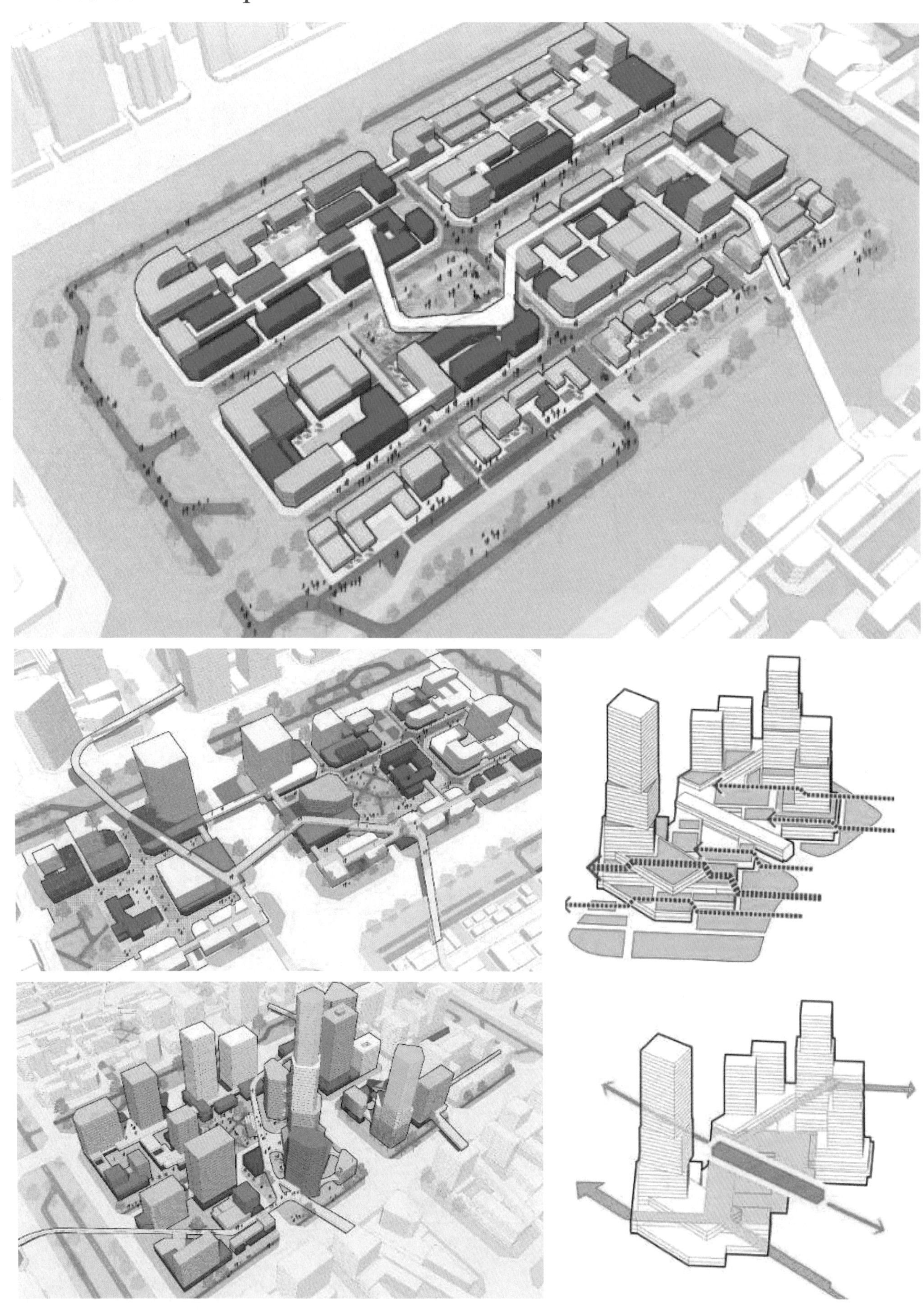

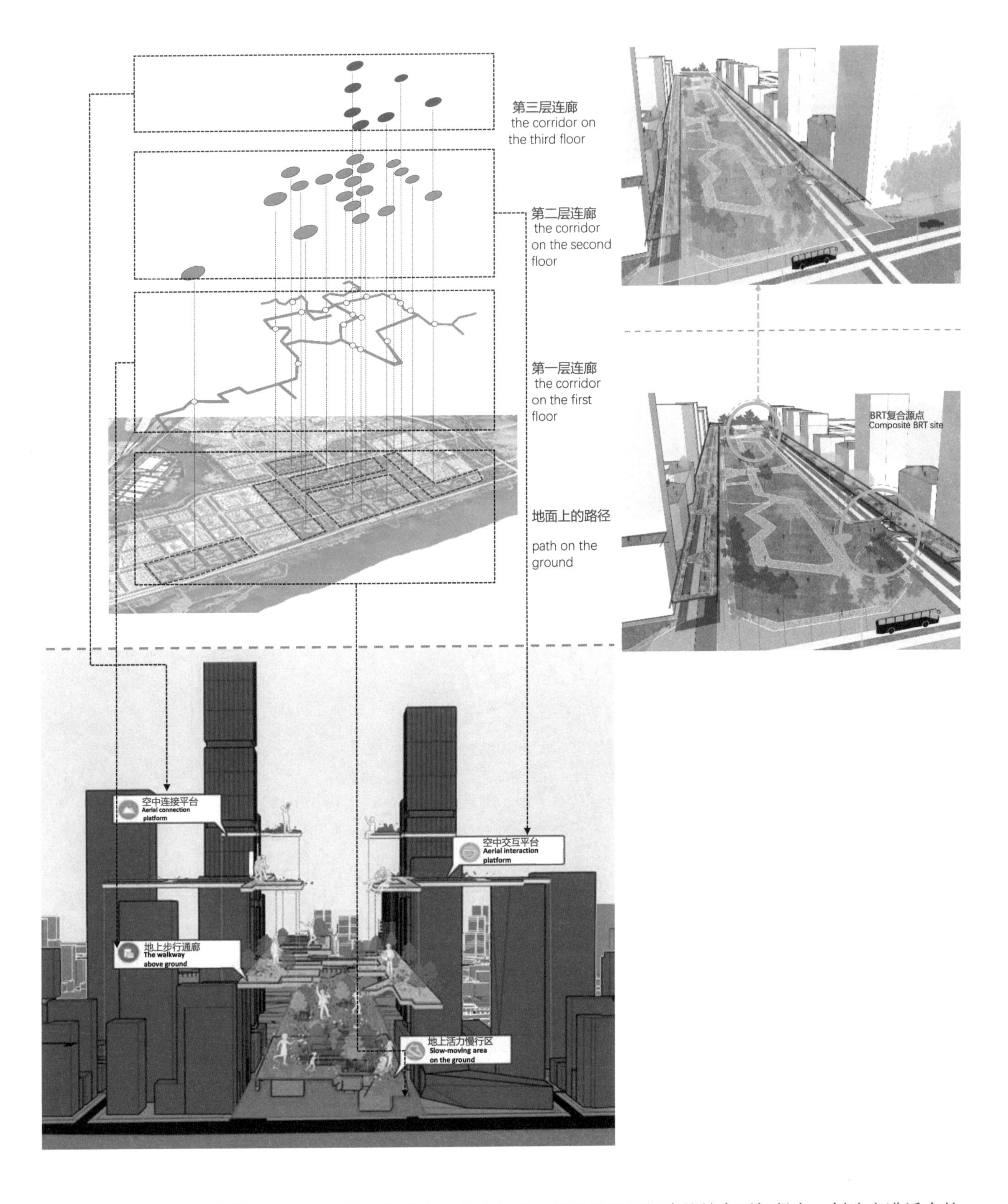

通过引入绿地与水体，结合灵活的建筑形态与布置方式，使蓝绿空间与建筑结合更加紧密，创造充满活力的商业空间，实现商业建筑、绿地、水体与活动空间的无界限融合。

Through the introduction of green space and water, combined with flexible architectural form and layout, the blue and green space is more closely combined with the building, creating a vibrant commercial space, and realizing the boundary-free integration of commercial buildings, green space, water and activity space.

总平面图 Master Plan

① 滨江湿地综合景观
integrated landscape of riverside wetland

② 滨江阳光草滩
riverside sunshine grass beach

③ 滨江系列观景平台
riverside series viewing platform

④ 空中云廊
air cloud corridor

⑤ 滨江亲水游步道
waterfront promenade

⑥ 府河生态绿地
Fu River ecological green space

⑦ 朱家河湿地
Zhu Jia River wetland

⑧ 朱家河公园
Zhu Jia River Park

⑨ 朱家河防洪景观平台
Zhu Jia River seasonal flooding landscape platform

⑩ 朱家河科技水舞剧场
Zhu Jia River technology water dance theater

⑪ 中央商业区——立体城市客厅
central business district—three-dimensional city living room

⑫ 滨江商务区——城市阳台
riverside business district—city balcony

⑬ 智慧核心社区
smart core community

⑭ 现代体育中心
modern sports center

⑮ 绿色医疗康养中心
green health care center

⑯ 中央公园——城市纽带
central park—urban bonds

⑰ 绿色宜居社区
green livable community

⑱ 滨江活力社区
vibrant riverside community

⑲ 智慧生活服务区
intelligent life service area

⑳ 慢行系统
slow system

㉑ 自动驾驶车道
autopilot lane

㉒ 创产研发中心
production r & d center

㉓ 科研交互中心
research interaction center

㉔ 创智孵化中心
creative intelligence incubation center

专项设计 Design of Special Topic

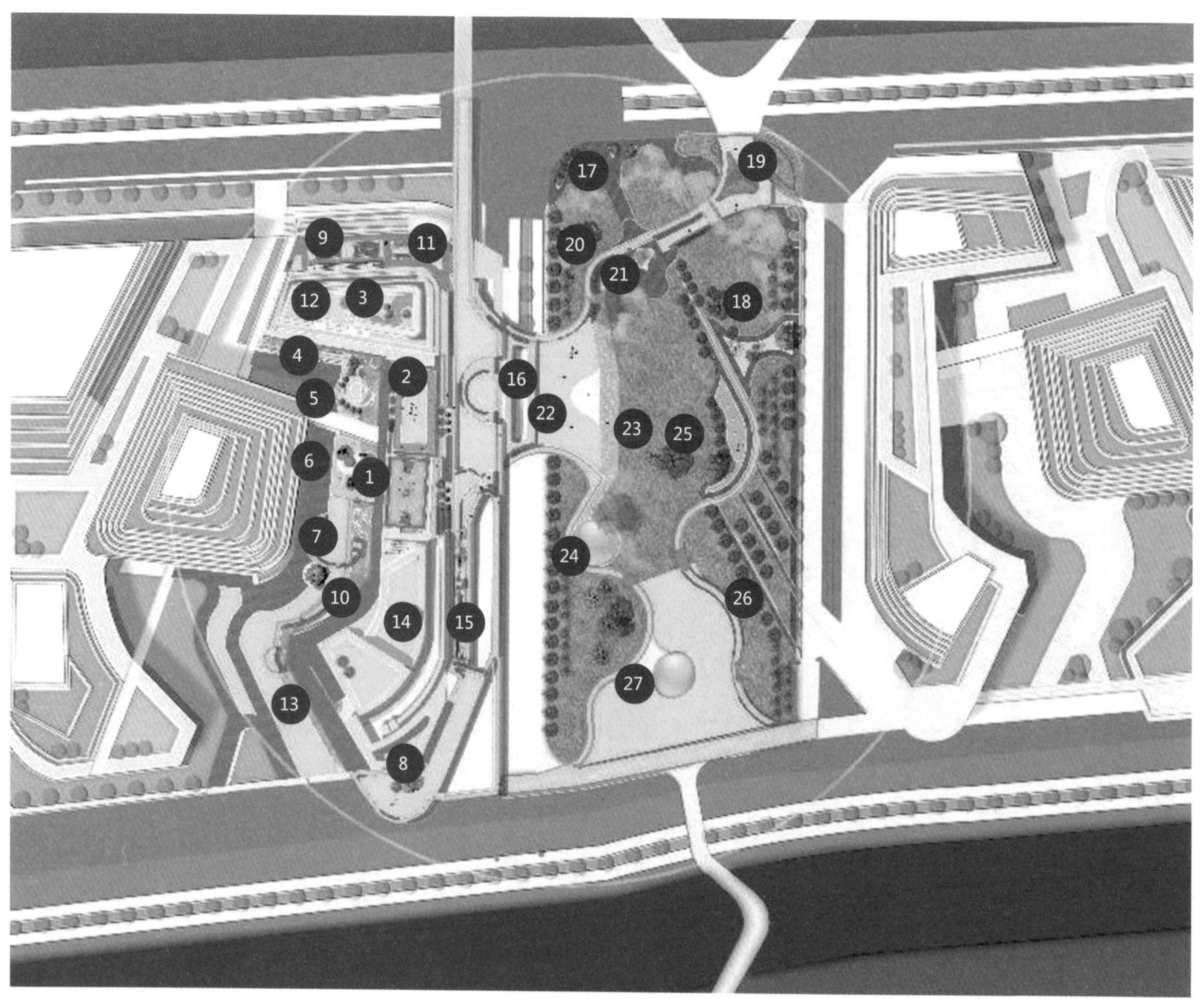

① 活力篮球场 energetic basketball court
② 露天微下沉舞台 open sunken stage
③ 屋顶阳光花园 roof sunshine garden
④ 动感水幕 dynamic water curtain
⑤ 沁泉广场 comfortable fountain square
⑥ 活动性艺术广场 active art square
⑦ 亲水平台 hydrophilic platform
⑧ 滨江观景台 riverside observation deck
⑨ 活力羽毛球场 badminton court
⑩ 凉亭小筑 pavilion
⑪ 小憩花园 leisure garden
⑫ 露台冥想花园 meditation terrace garden
⑬ 水剧场草坪 water theater lawn
⑭ 绿色阳台 green balcony
⑮ 漫游植筑 vertical green buildings
⑯ BRT中心 BRT center
⑰ 入口广场 entrance square
⑱ 花境 flower border
⑲ 绿丘 green hill
⑳ 律动天桥 rhythm bridge
㉑ 花花广场 flower square
㉒ 趣味滑草 gout grass skiing
㉓ 阳光草坪 sun lawn
㉔ 旱喷广场 dry spray square
㉕ 采光井 light well
㉖ 树阵广场 tree array square
㉗ 下沉星河广场 sunken star square

该区域整合了不同的活动功能区和多种景观元素，以创建一个满足沟通、休闲和其他日常工作多功能需求的综合环境。

Different functional areas and landscape elements are integrated to create a comprehensive environment to meet the need of communication, leisure and other daily work.

建筑推衍 Building Deduction

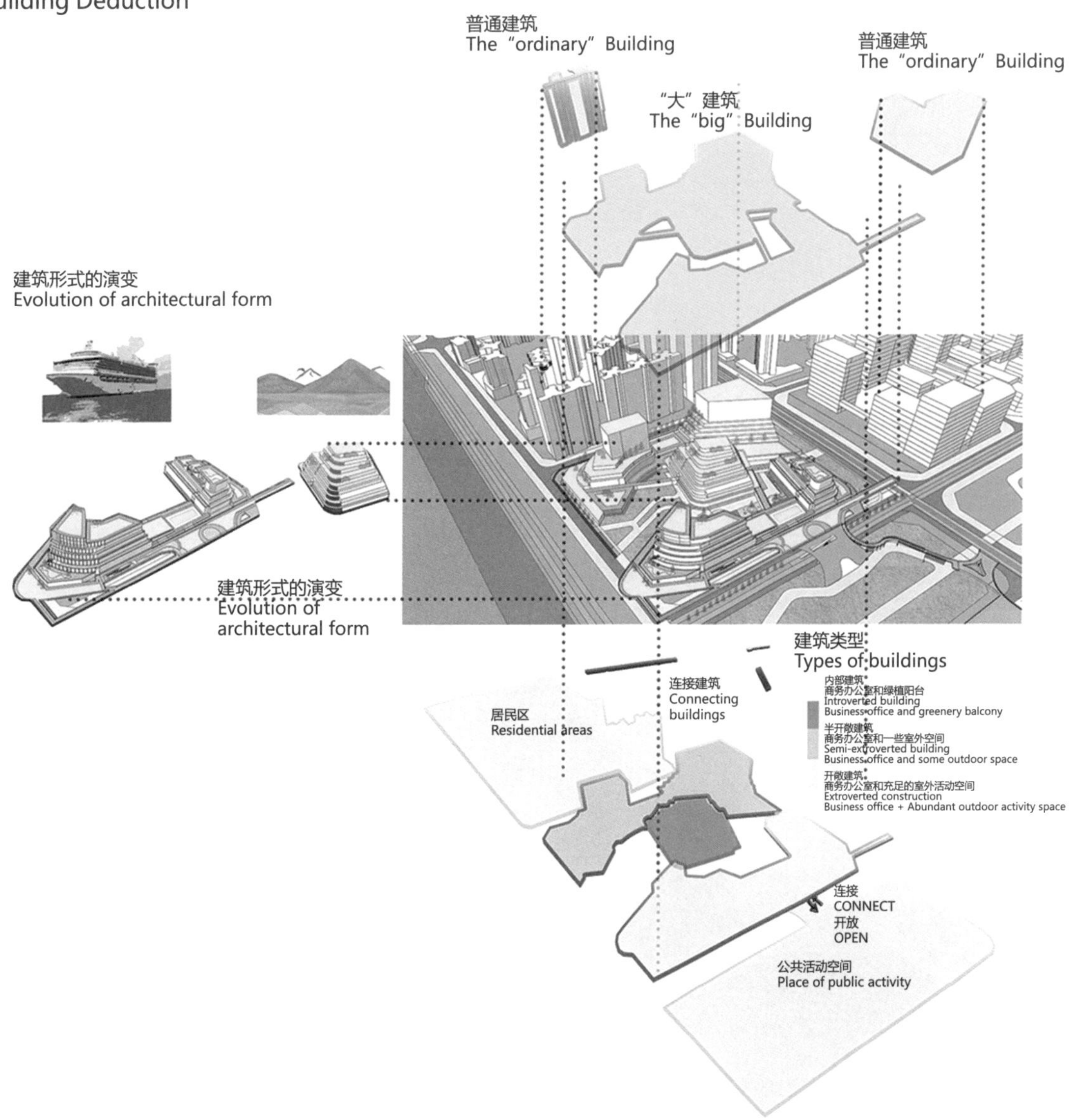

节点分析 Landscape Node Analysis

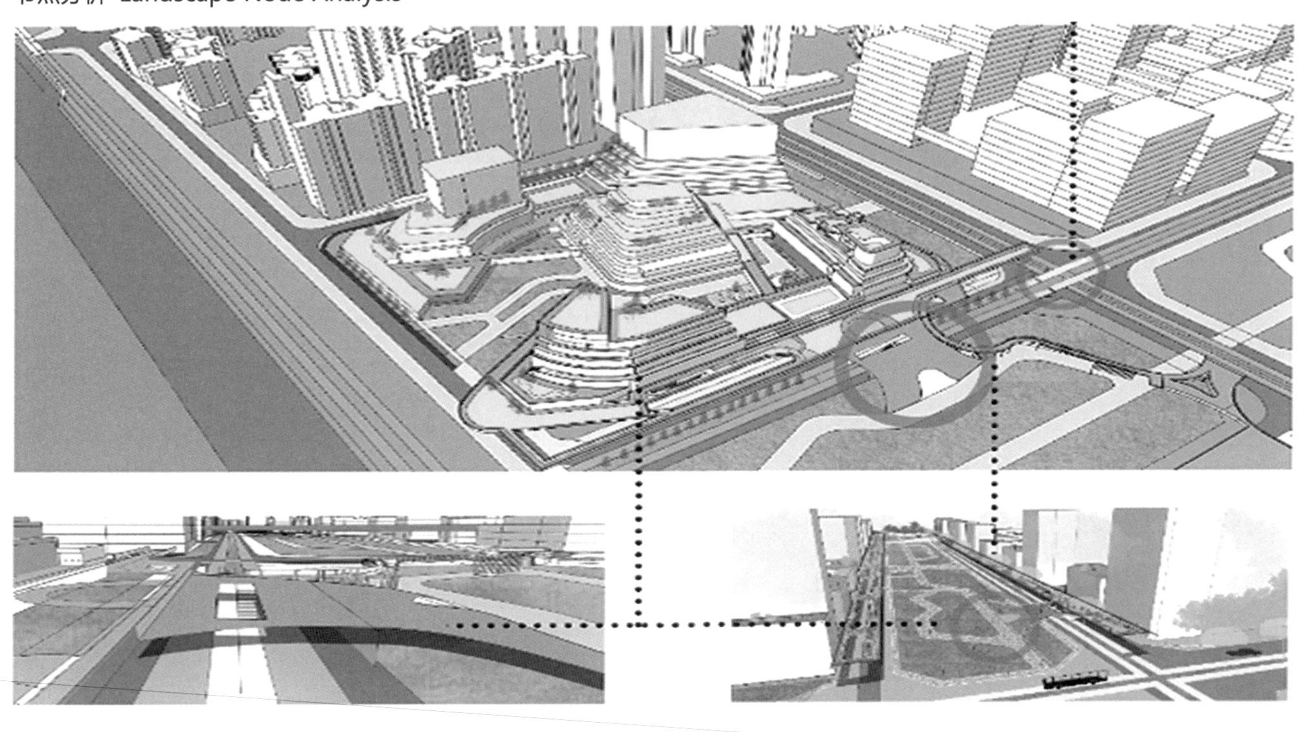

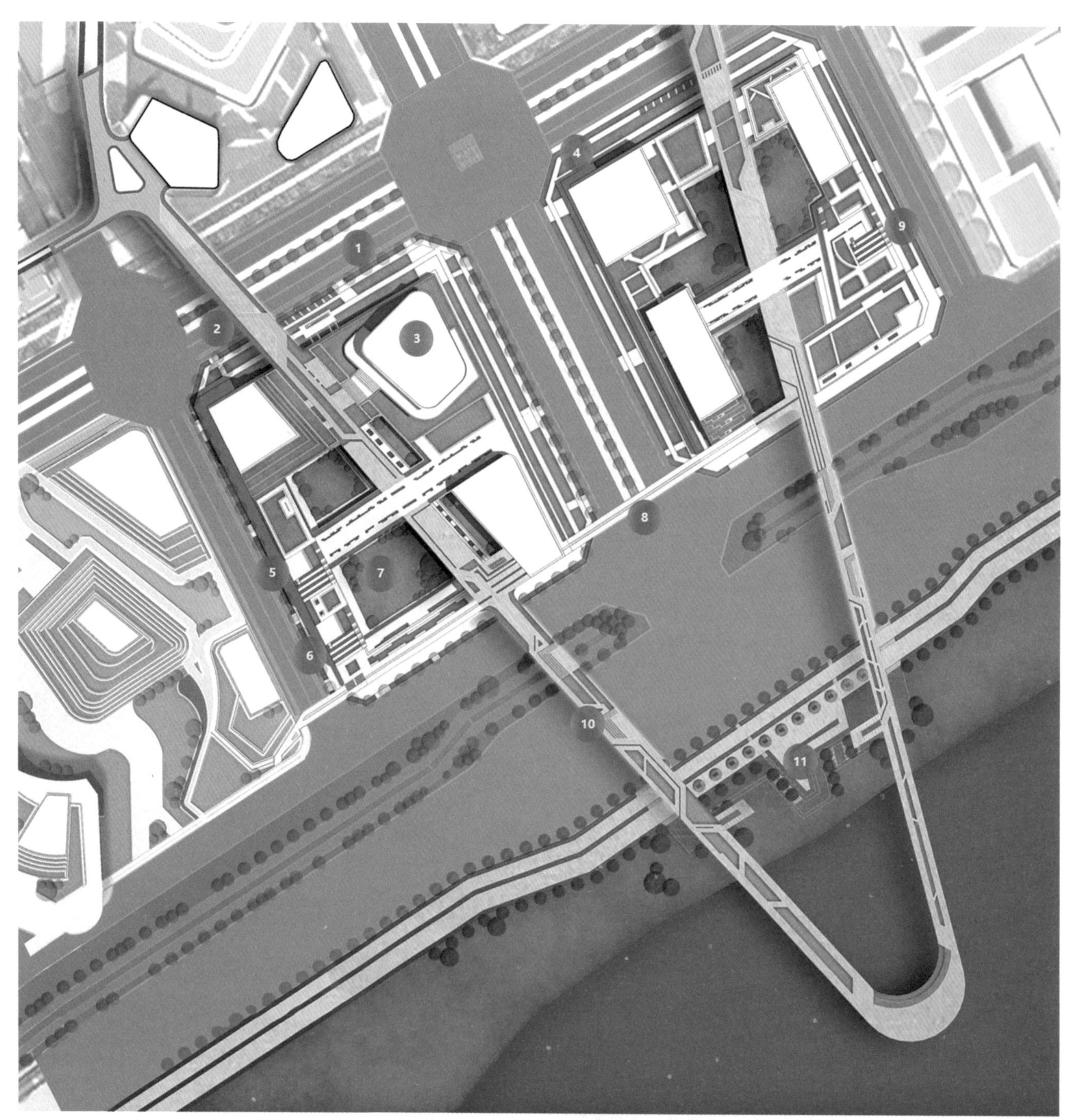

① 街道公园 roadside park
② 骑行道，自行车停车区 cycling lanes, bike parking areas
③ 商务办公楼 commercial-office buildings
④ 一二层商业 first and second floor commerce
⑤ 秘密花园 secret garden
⑥ 台阶园艺池 stepped garden pool
⑦ 底层中庭 lowest atrium
⑧ 过街绿廊 green corridors across the street
⑨ 天台花园 rooftop garden
⑩ V形看台 V-shaped viewing platform
⑪ 滨江广场，栈道 riverside square, plank road

多个绿色空间散落其中，连廊在特定的绿色开放空间设置落地设施，使居民随时可通过绿色连廊进入绿色空间，实现商务建筑与绿地的无界化。

Residents can enter the green spaces through the green corridors, which are scattered among in and are equipped with floor-to-ceiling facilities at specific time, realizing the unboundedness of commercial buildings and green spaces.

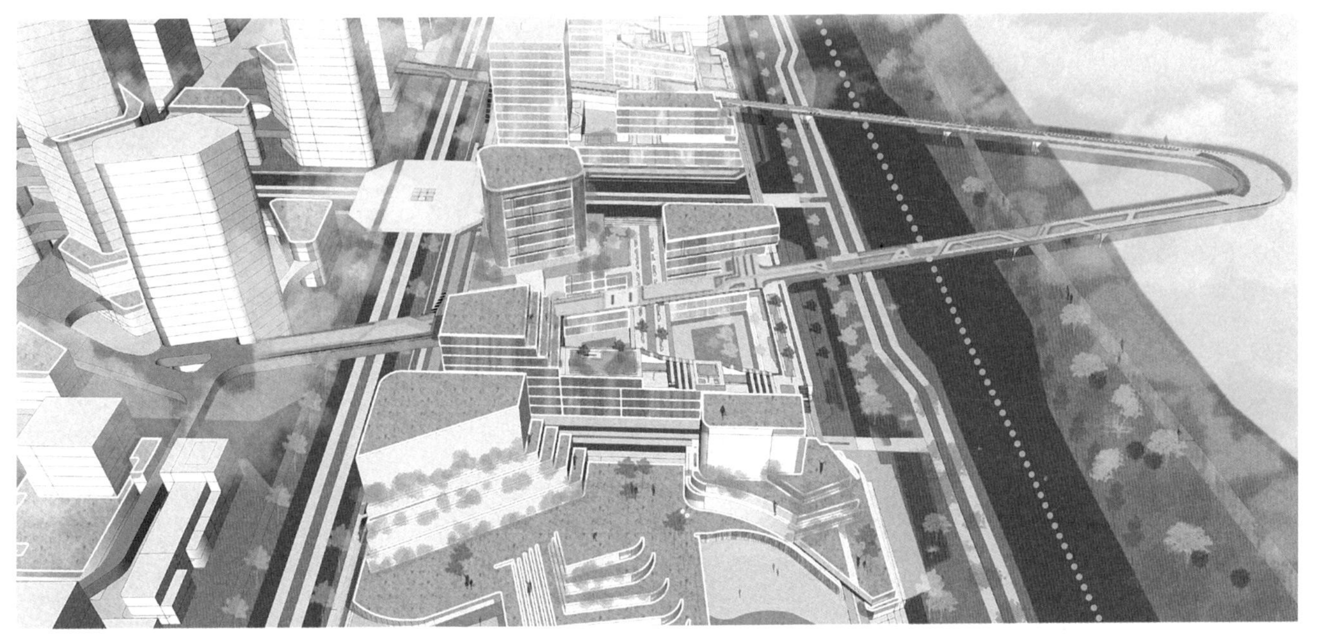

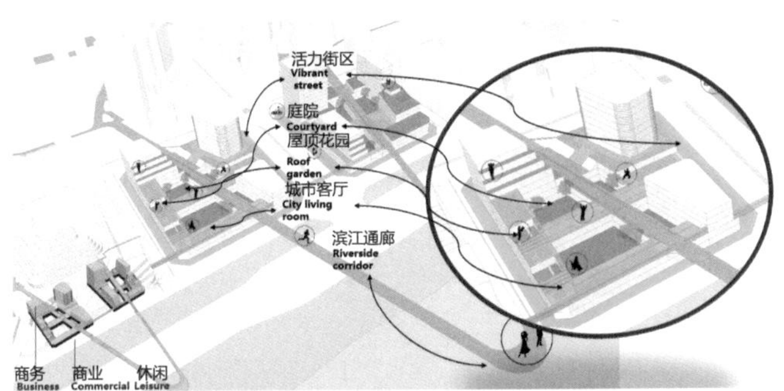

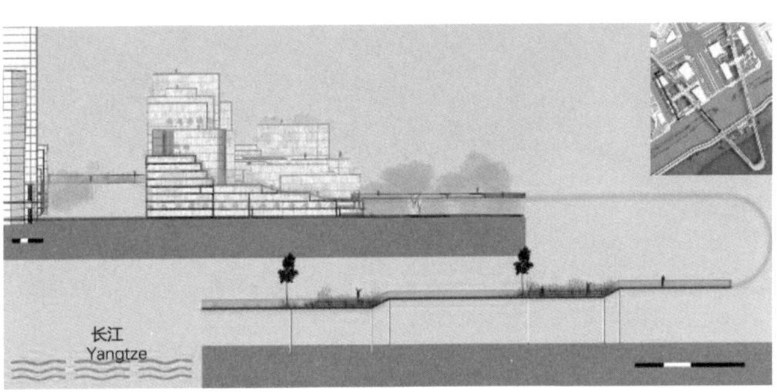

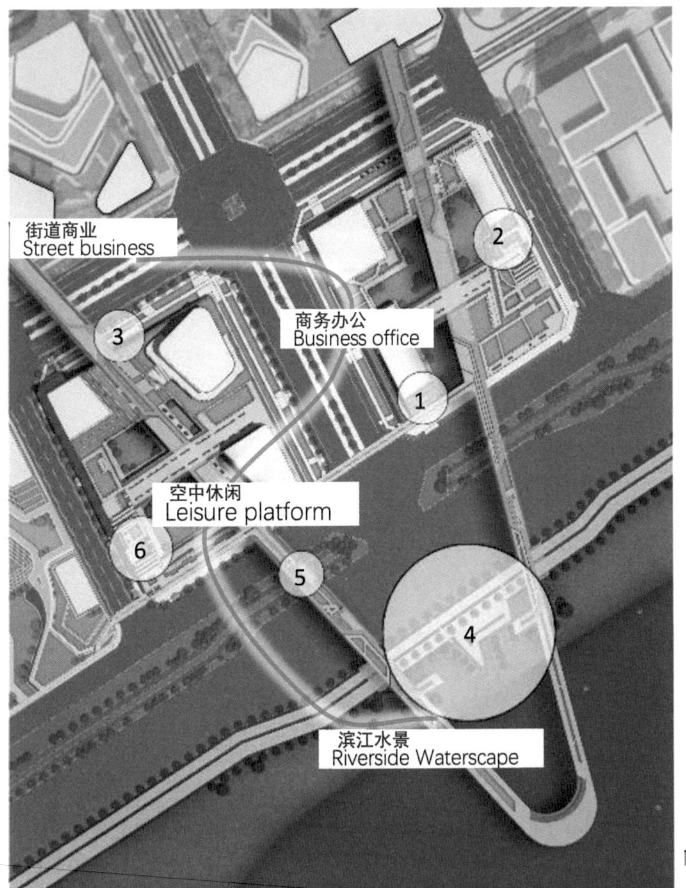

① 商务附属-休闲阳台
Business: Leisure balcony

④ 滨江水景江滩广场
Riverside:
Riverside Plaza

② 商务附属-双层天台花园
Business: Double roof garden

⑤ 空中休闲-台阶观景台
Leisure: Ladder Observation Deck

③ 街道商业-休闲广场
Street：
Waterscape Plaza

⑥ 空中休闲-互动种植台
Leisure: Interact planting platform

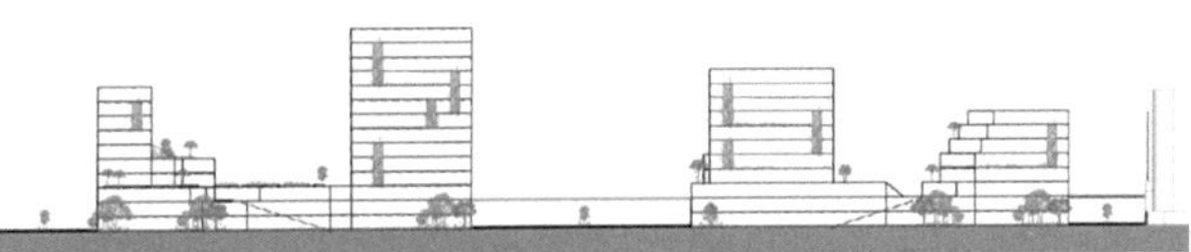

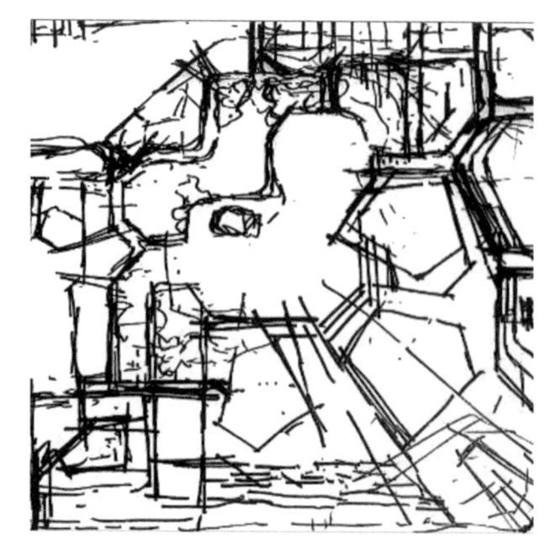

① 会展中心
convention center

② 博物馆
museum

③ 市民文化中心
civic cultural center

④ 入口广场
entrance square

⑤ 临水亭
waterfront pavilion

⑥ 下沉花园
sunken garden

⑦ 特色商业区
characteristic business district

⑧ 亲水草坪
waterfront lawn

⑨ 林间小道
forest trail

⑩ 喷泉
fountain

⑪ 亲水平台
hydrophilic platform

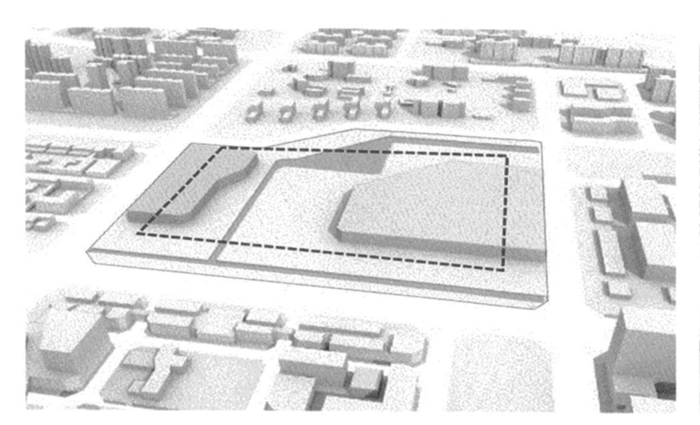

联系商业用地与公共服务设施用地
Contact commercial land with public service land

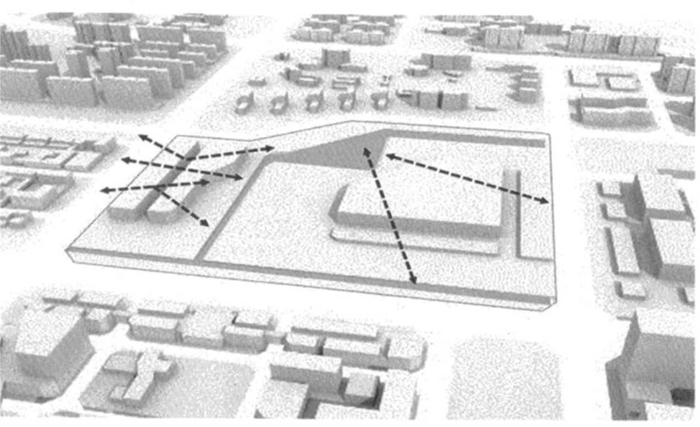

打破建筑与绿地、水体之间的分隔
Break the separation between buildings, green spaces and water

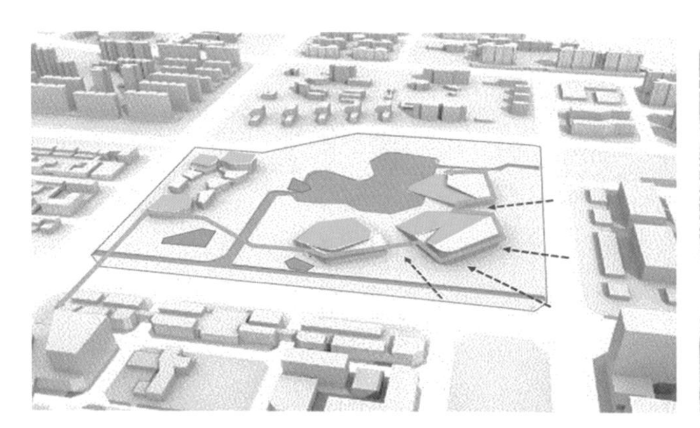

整体交通及景观的细化提升
Overall traffic and landscape refinement

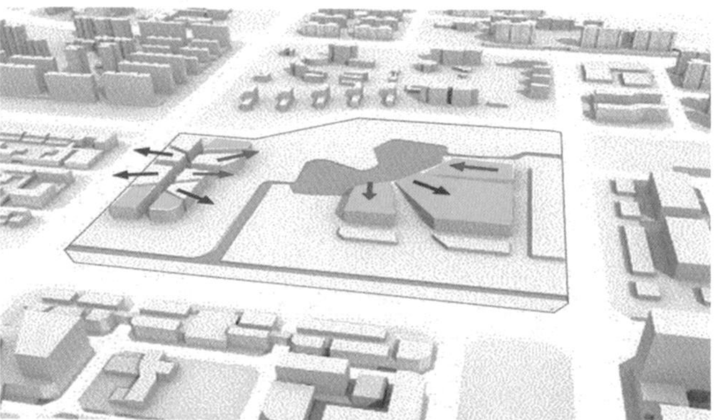

形成半围合的公共开放空间
Forming a semi-enclosed public open space

A-A 剖面图
A-A' section

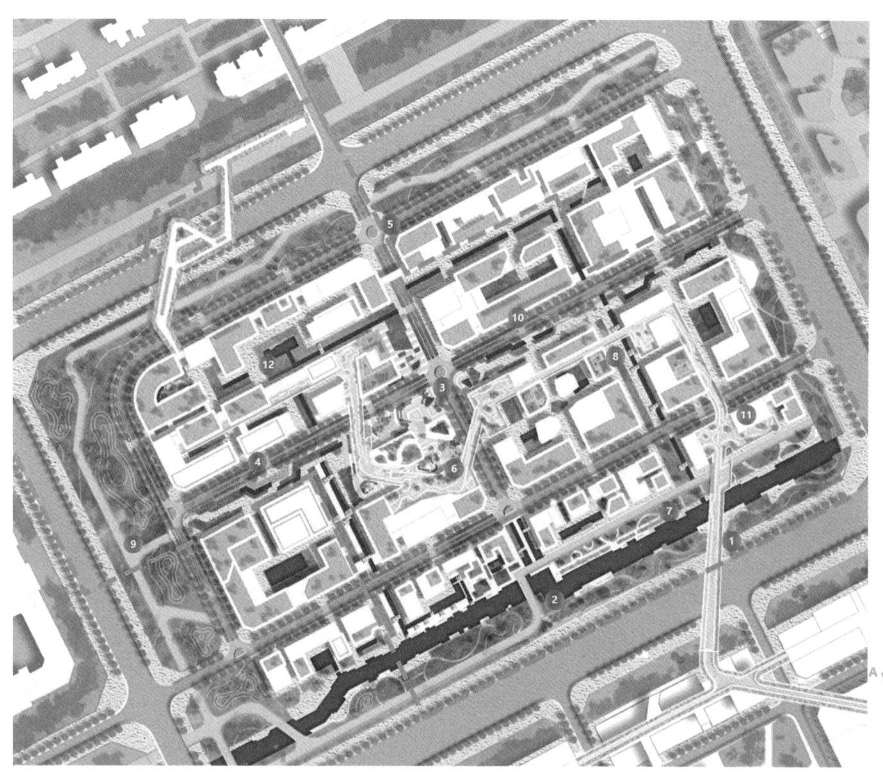

① 慢行云廊
slow corridor

② 无界水廊
the unbounded water gallery

③ 中央活动广场
central event square

④ 区域内慢行系统
slow traffic system in the area

⑤ 互动休憩亭
interactive rest pavilion

⑥ 覆土交互建筑
interactive earth-covered building

⑦ 滨水商业街
waterfront commercial street

⑧ 绿色交互平台
green interactive platform

⑨ 休憩游步道
recreation trail

⑩ 休闲步行街
leisure pedestrian street

⑪ 屋顶花园
roof garden

⑫ 互动水渠
interactive canals

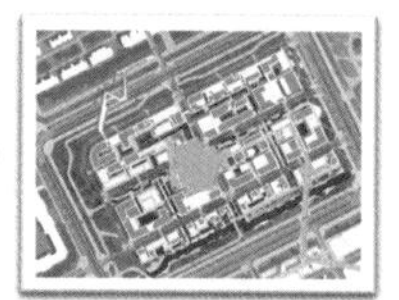

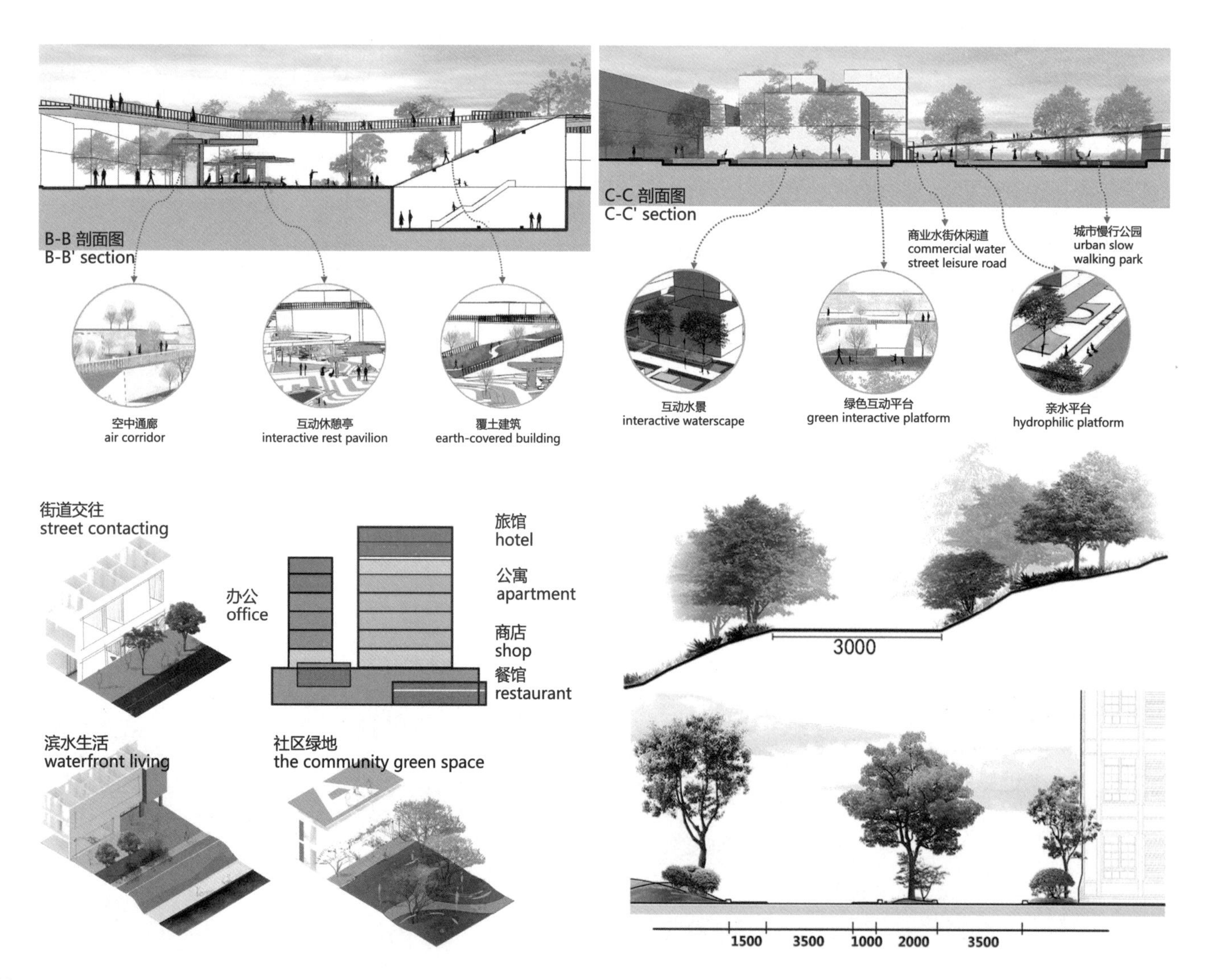

空层、中庭、开放的街区环境、良好的景观等共同构成了一个促进交流的互动空间，是打破生活界限的重要触媒。

The empty floor, the atrium, the open block environment, and the good landscape together constitute an interactive space that promotes communication and is an important catalyst for breaking the boundaries of life.

开阔而美丽的朱家河水湾如同蓝色的山岬嵌入街区，而开阔的视线通廊充分利用了这一良好景观。

The wide and beautiful Zhujia River Bay is like a blue cape embedded into the block. And the wide view corridor fully realizes the maximum utilization of this good landscape.

详细设计 Detailed Design

第二部分　美国学生作品

Works from America

滠水湿地廊道

Sheshui Wetland Corridor

Hannah Slyce

滠水河作为长江的主要支流，其与长江汇聚之处对长江新城的生态及城市健康尤为重要。此公园作为湿地走廊，在提供绿色雨水基础设施的同时，还构建了一个 ± 10km 的多式联运网络以连接多个地块，为居民提供具备娱乐、教育功能的都市空间。这一线性走廊分为两个区域，每个区域有不同主题及规划要素。而整个场地将重点关注三大要素——城市森林、多式联运和水文，具体的子区域包括公园空间、公共空间以及自然恢复区。

With the Sheshui River at its apex and a major tributary to the Yangtze River at its cusp, this site is extremely important to the ecology and urban health of the New City. This park space serves not only as a wetland corridor and green stormwater infrastructure, but also as a ± 10 kilometer multi-modal transportation network that connects districts that provides a recreational and education urban space for residents. The linear corridor will be broken into 2 sub-areas with each different programmatic elements and themes. While the entire site will focus on three major elements, urban forestry, multi-modal transportation, and hydrology, the specific sub-areas contain park spaces and public realms along with natural restorative areas.

节点 KEY

① 滠水岸线修复 Sheshui River edge restoration

② 城市森林网络 urban forest network

③ 连续的水文系统 continuous hydrological system

④ 公园空间 park space

⑤ 河岸码头及瞭望台 lake pier and lookout

⑥ 雨水滞留池 stormwater retention pond

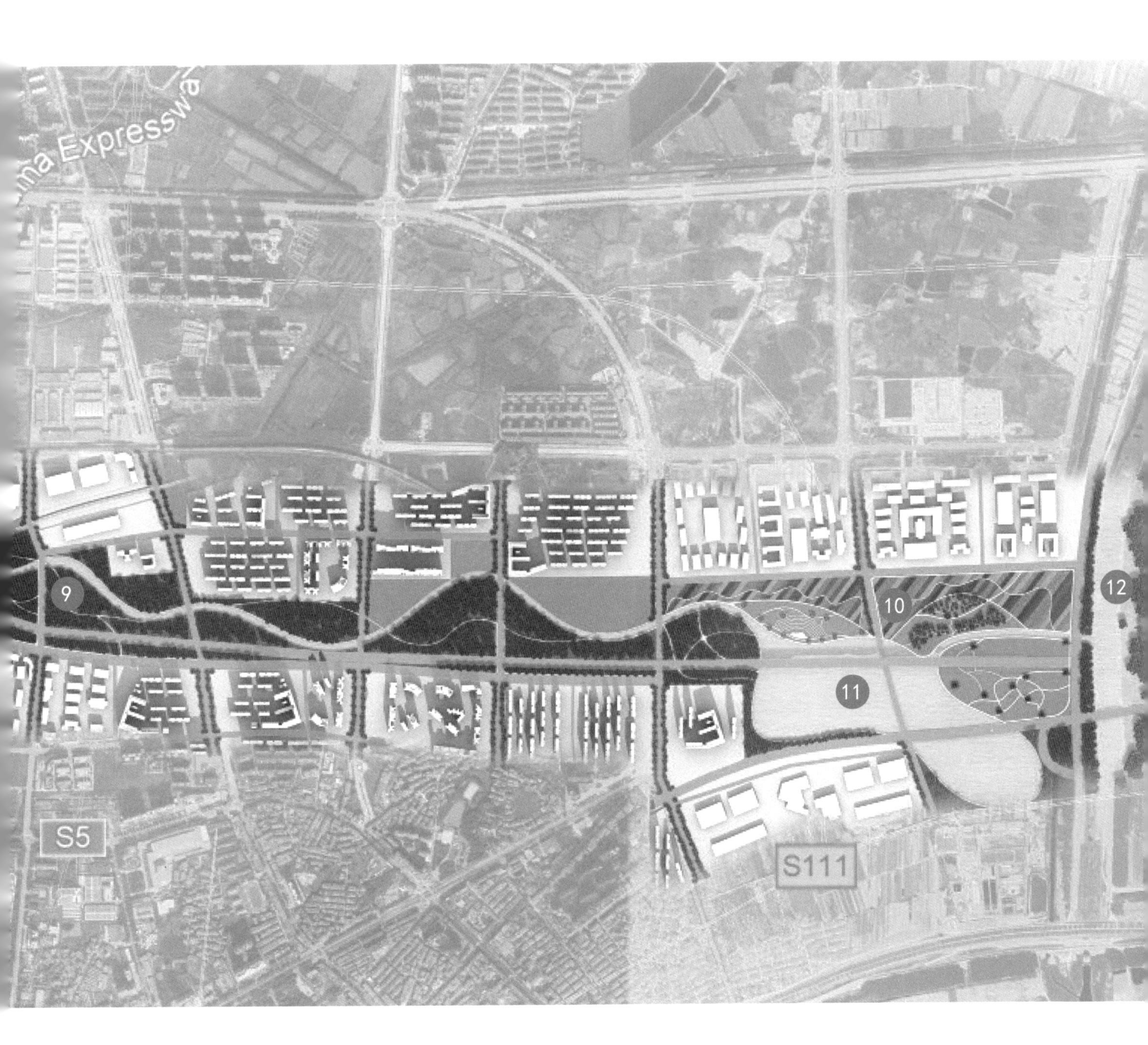

⑦ 连续的城市森林及河流 continuous urban forest & stream

⑧ 扩建的医用池塘 expanded pond for hospital usage

⑨ 密集型人行交通系统 intensive pedestrian path system

⑩ 农场公园 agricultural park

⑪ 扩建的历史稻田湖 expanded historical rice paddy-turned lake

⑫ 支流修复及其优势 tributary restoration & advantage

设计策略 Design Strategy

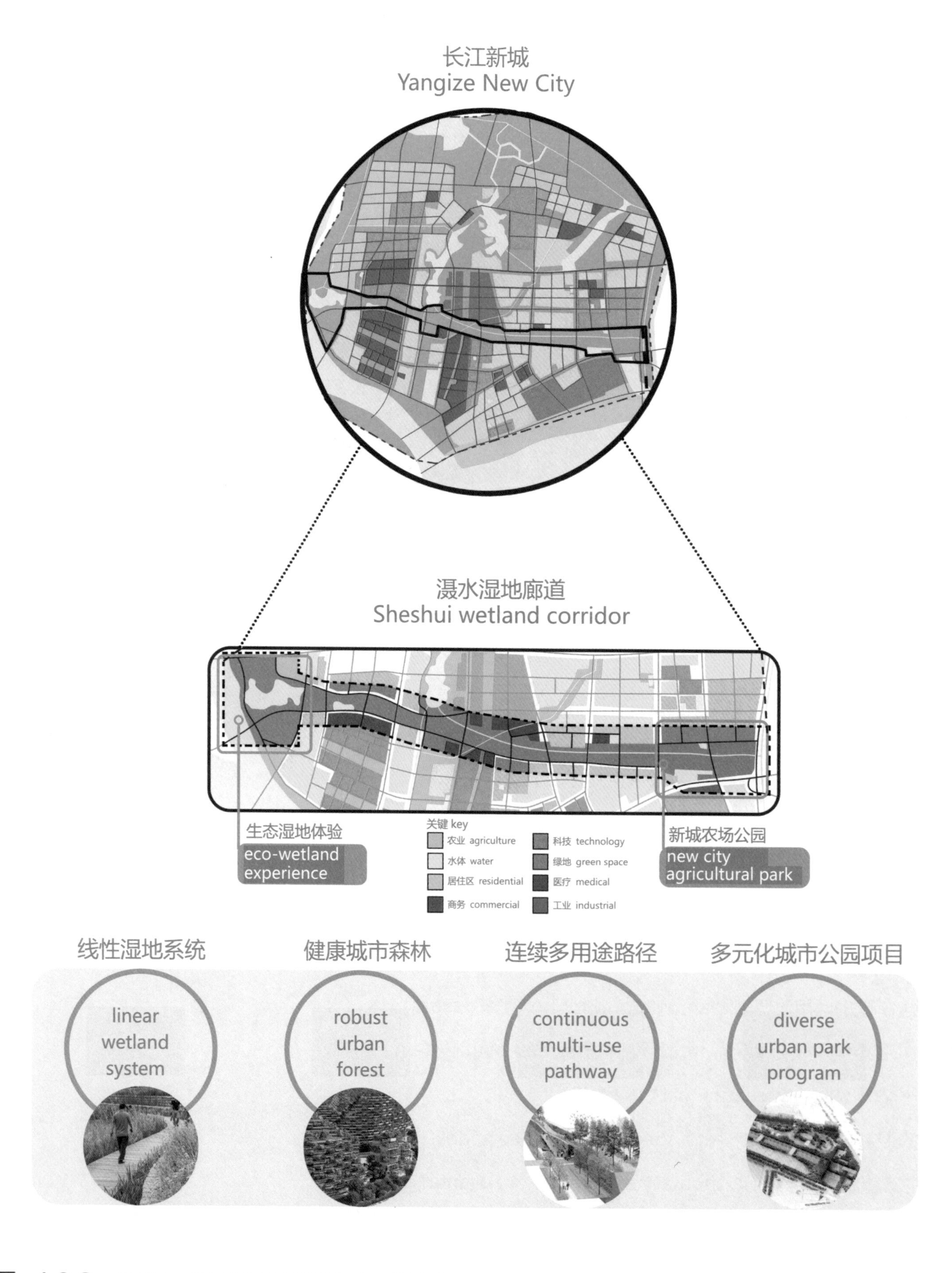

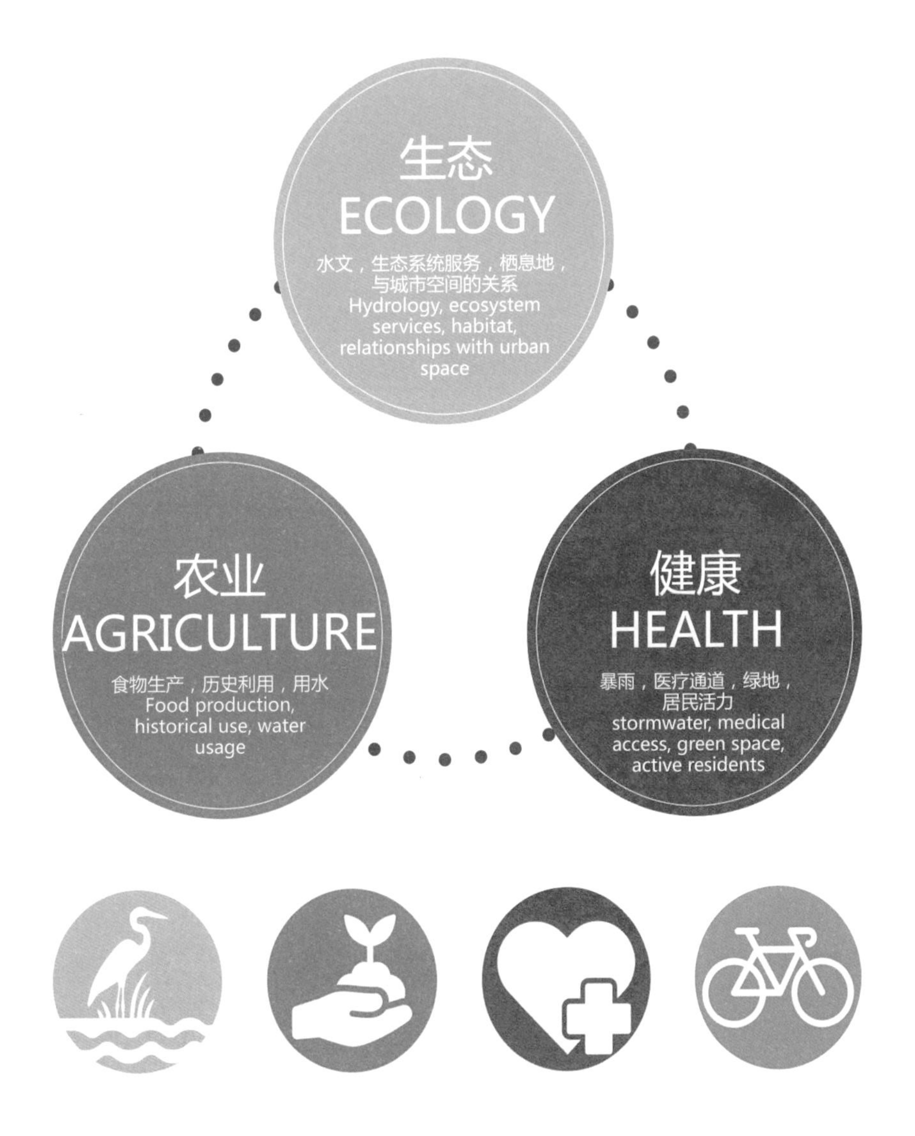

以上三个设计原则都根植于滠水湿地廊道项目中。生态是该方案的灵魂，场地内所有生物的生态都被纳入考虑范畴。从湿地到水文再到城市生活，都是长江新城生态的一部分。农业作为场地的历史用途，是该方案的核心，而缅怀过去则是主要的设计理念。利用原有场地用途建立都市农场以达到寓教于乐的目的是设计的重点，并可借机进行粮食储备和用水控制。最后，健康是将一切联系在一起的纽带，毕竟健康的生态系统和健康的农业实践都意味着这是一个健康的城市。湿地廊道线型布局将促进交通活力，有助于空气流通，也为周边提供了雨水补给。

Each of these three design principals are ingrained into the Sheshui Wetland Corridor project. Ecology is seen as the soul of the proposal, considering the ecology of all living things within the site, from wetlands to hydrology to urban life. All things are part of the ecology of the Yangtze River New City. Agriculture comes next as the heart of the proposal. As the historical use of much of the site, remembering the past is a major design concept. Educating the public on this historic use through urban agriculture will be a focus along with food readiness and proper water usage. Lastly, health will be the bond that everything together, as a healthy ecosystem and healthy agriculture practices means a healthy city. The linear aspect of the corridor will encourage active transportation, allow for fresh air flow, and allowing for a stormwater recharge zone for the adjacent urban districts.

生态湿地体验
Eco-Wetland Experience

节点 Key

① 樱花步道 cherry blossom path

② 步道/跑道 walk/run path

③ 荷塘和四季苗圃 lotus ponds & perennial beds

④ 小土坡和儿童游乐区 hill & children's play area

⑤ 参观教育中心 visitors & education center

⑥ 湿地露天剧场 wetland front amphitheater

新城农场公园
New City Agricultural Park

节点 Key

① 农耕体验步道 agricultural experience trail

② 湖岸露天剧场 lake front amphitheater

③ 历史稻田湖 historical rice-paddy pond

④ 都市农场 urban farm

⑤ 都市果园 urban fruit orchard

⑥ 开放空间 open space

详细设计 Detailed Design

土地利用分区
Diagram of Land Use

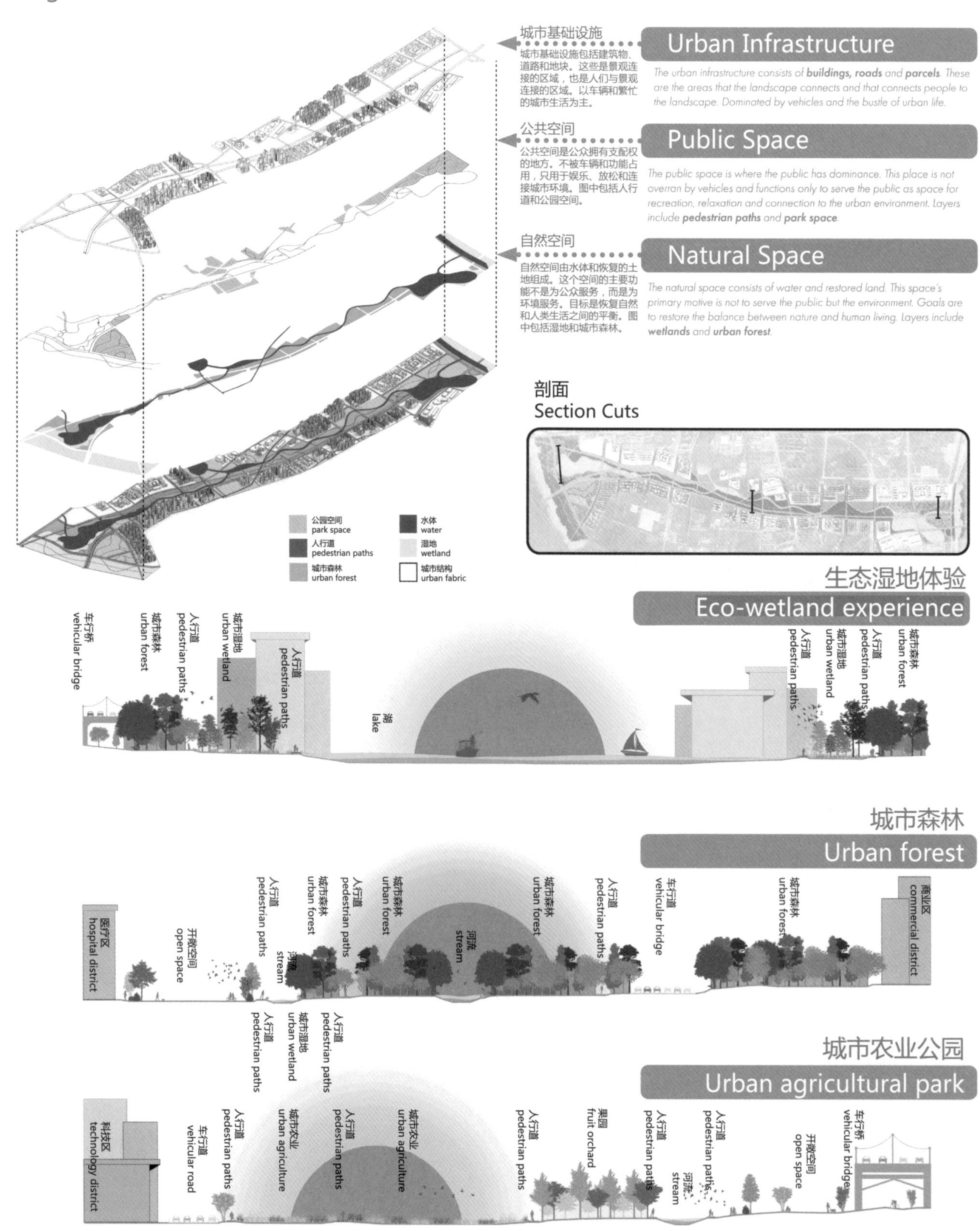

武汉医疗区
—— 生态疗养城
Wuhan Medical District
—The Healing Eco City

XIN Lyu

本方案对新型综合医疗区进行开发和提升，其中包括医疗（医疗诊所）、医疗研究（医疗研究机构）、医疗产品制造（医疗私营公司）。此外，本设计将促进城市可持续发展及交通可达性；并建设绿色基础设施，解决雨洪问题，减少热岛效应。

Develop and enhance a new and comprehensive medical district area that comprise medical treatment (clinics), medical research (medical research institution) and medical product manufacturing (medical private companies). Besides, this design will promote urban sustainable development and accessibility of transportation. The building of green infrastructure will handle stormwater problem and reduce heat island effect.

整体鸟瞰图
AERIAL VIEW OF MASTER PLAN

设计策略 Design Strategy

区位图
location map

土地利用图
land use diagram

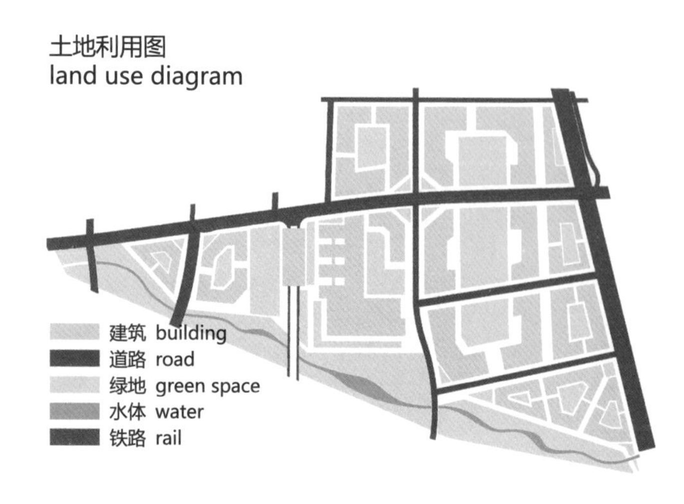

绿地系统图
green system diagram

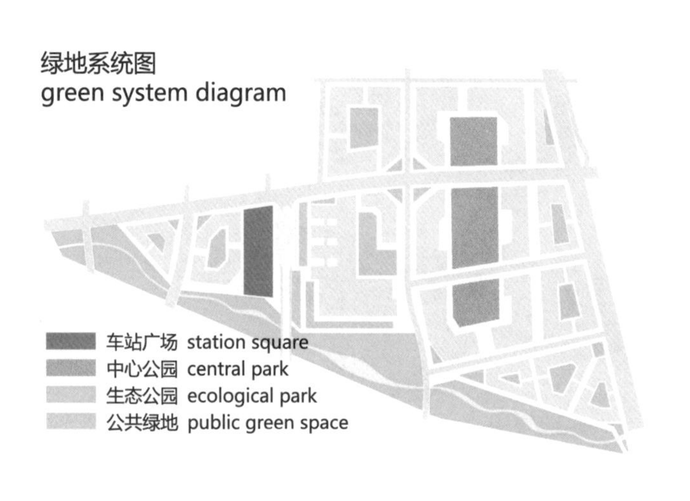

道路图
circulation diagram

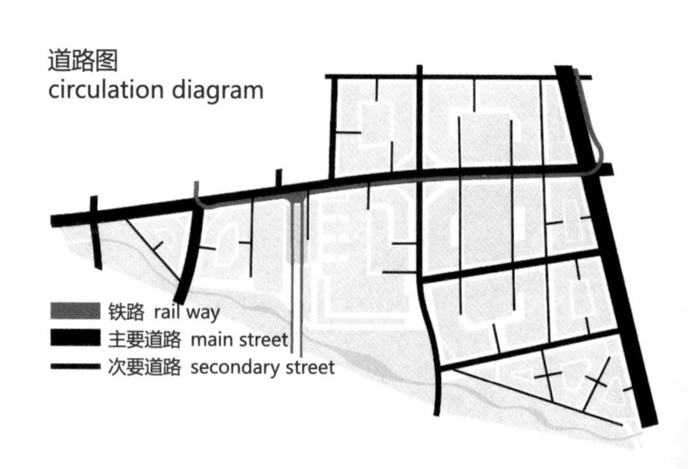

建筑利用图 building use diagram

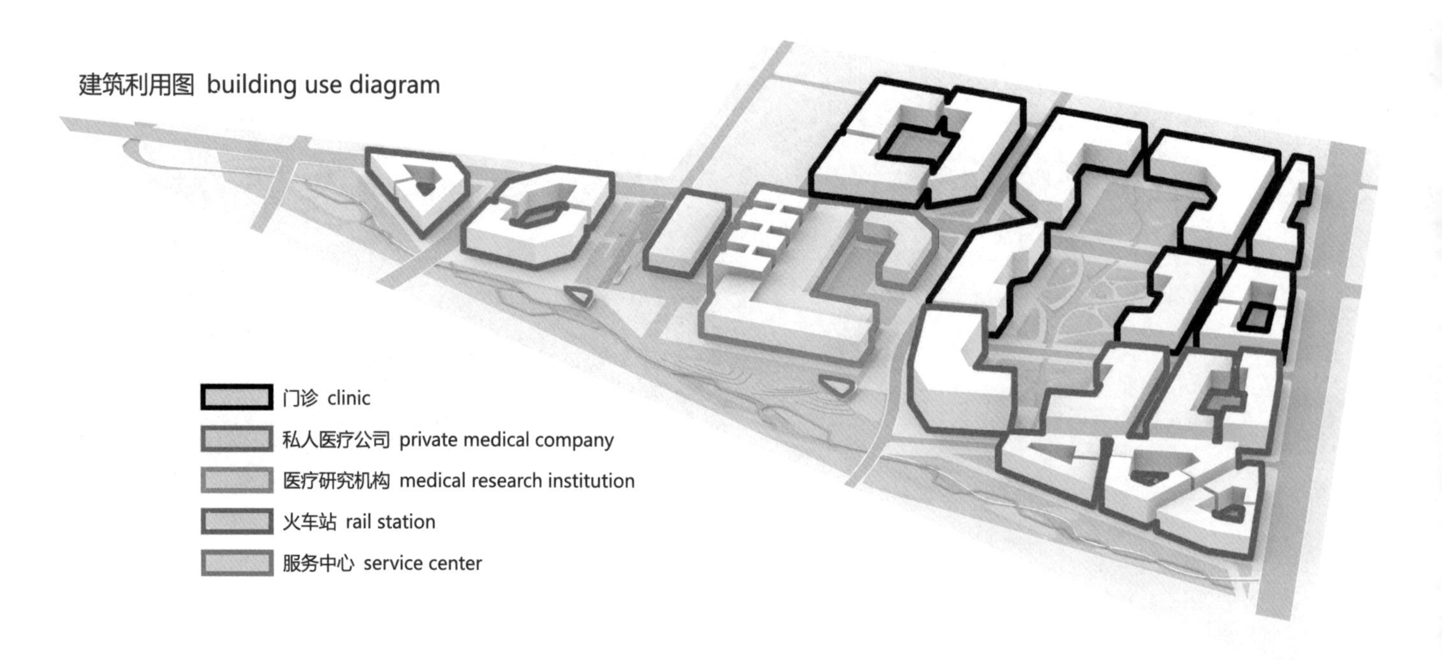

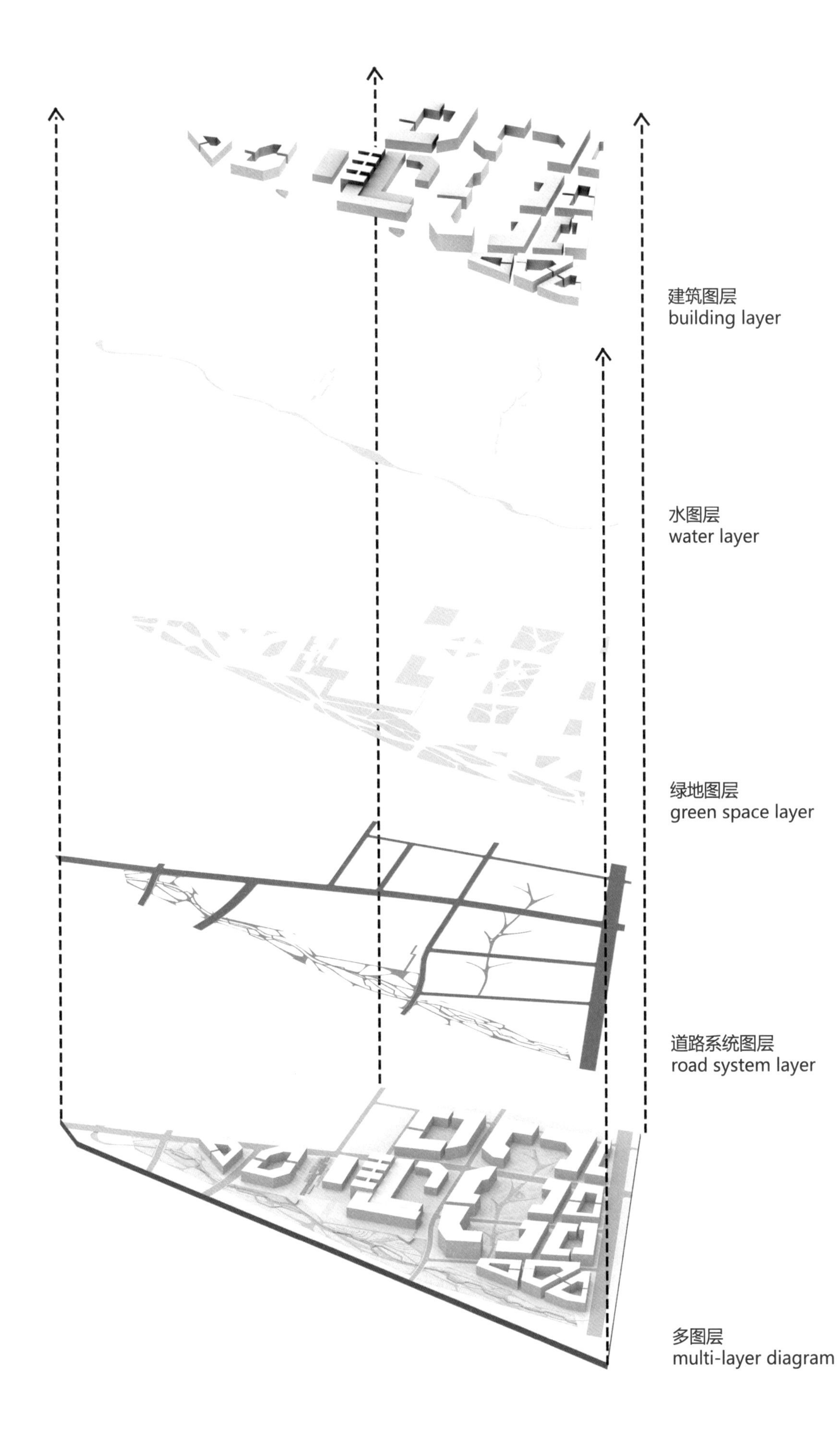
建筑图层
building layer
水图层
water layer
绿地图层
green space layer
道路系统图层
road system layer
多图层
multi-layer diagram

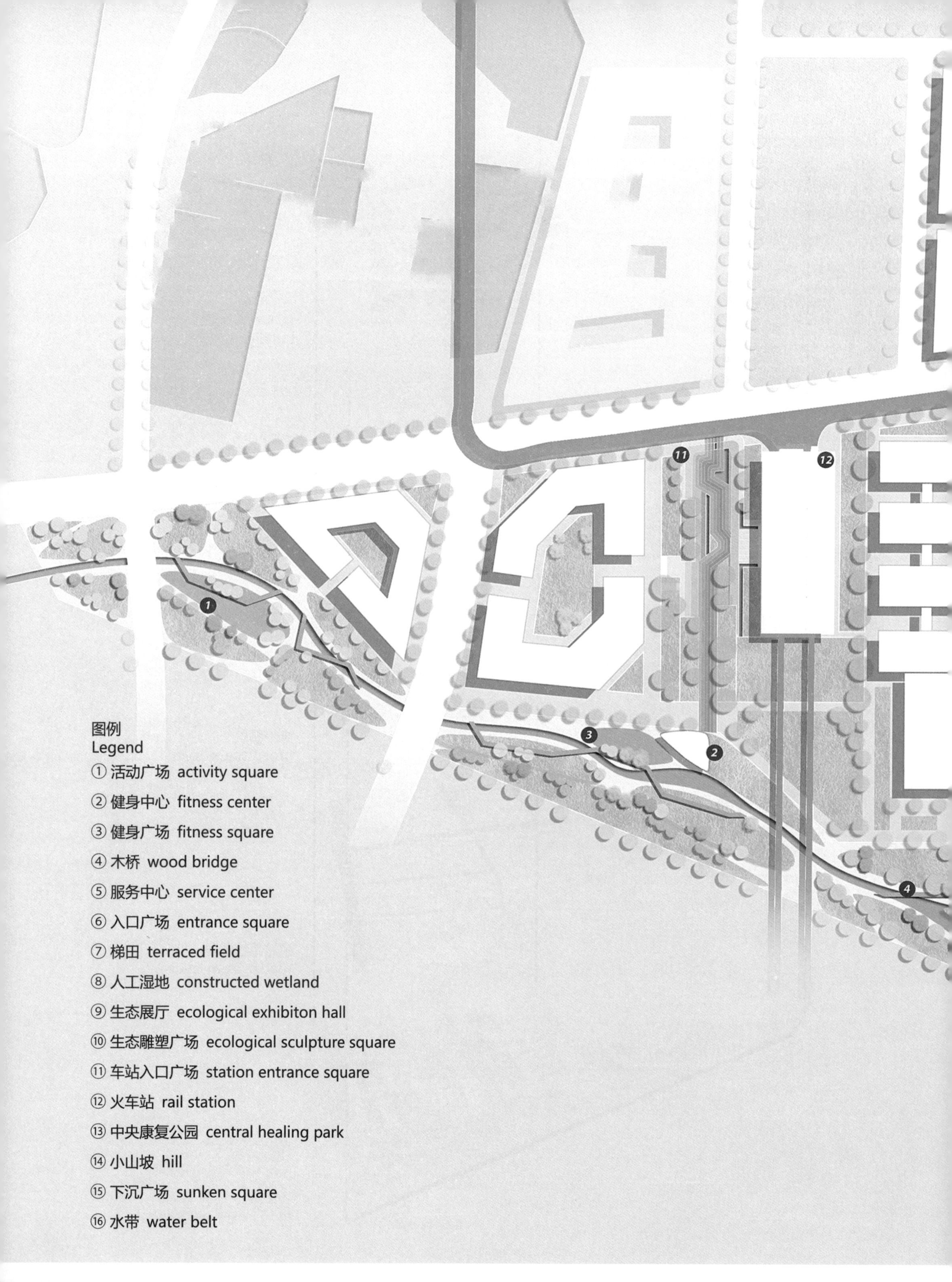
11
12
1
3
2
4
图例
Legend
① 活动广场 activity square
② 健身中心 fitness center
③ 健身广场 fitness square
④ 木桥 wood bridge
⑤ 服务中心 service center
⑥ 入口广场 entrance square
⑦ 梯田 terraced field
⑧ 人工湿地 constructed wetland
⑨ 生态展厅 ecological exhibiton hall
⑩ 生态雕塑广场 ecological sculpture square
⑪ 车站入口广场 station entrance square
⑫ 火车站 rail station
⑬ 中央康复公园 central healing park
⑭ 小山坡 hill
⑮ 下沉广场 sunken square
⑯ 水带 water belt

13
15
14
16
5
6
9
10

专项设计 Design of special topic

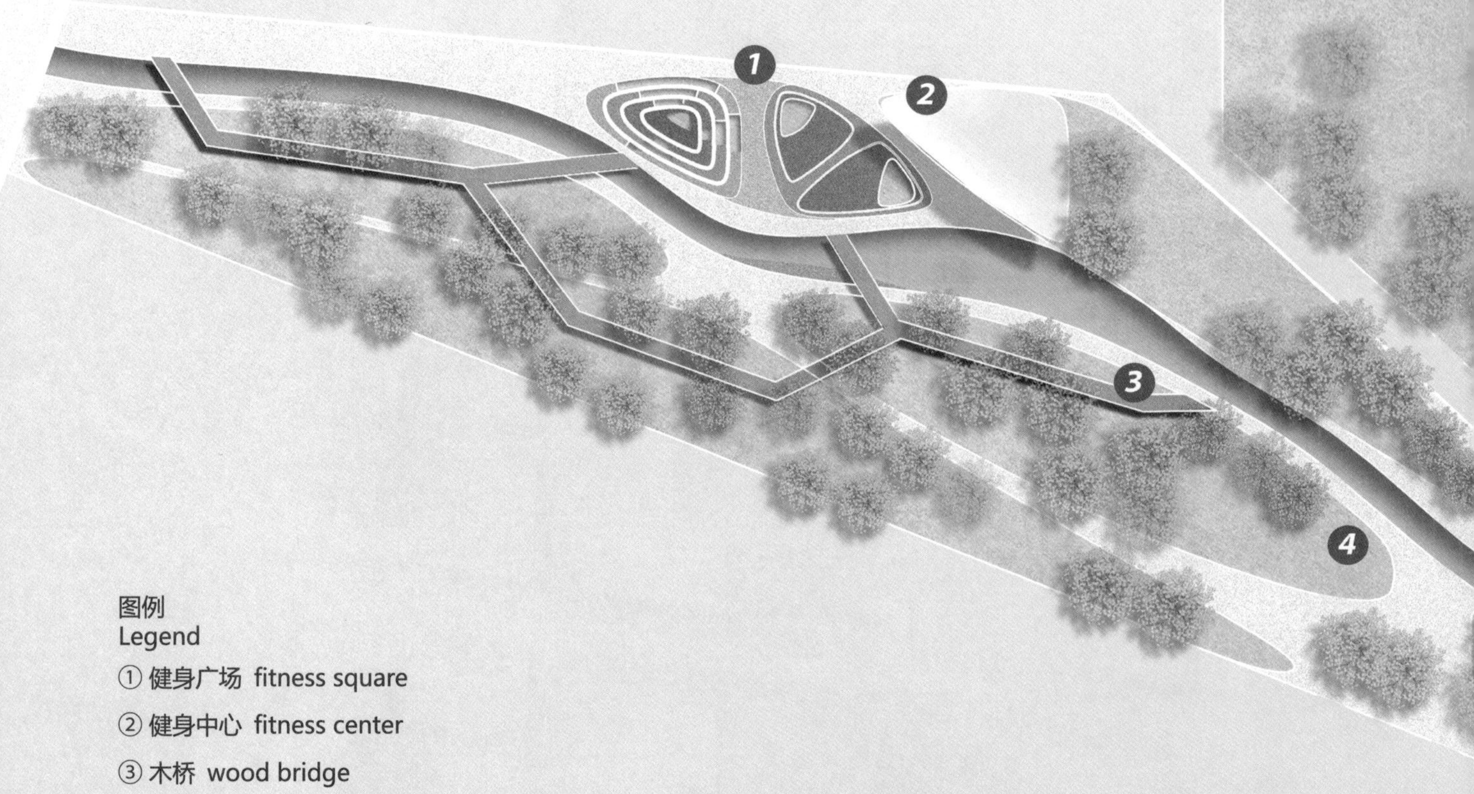

图例
Legend

① 健身广场 fitness square
② 健身中心 fitness center
③ 木桥 wood bridge
④ 植被区 vegetation area
⑤ 梯田景观 terraced landscape
⑥ 人工湿地 constructed wetland
⑦ 滨水广场 waterfront square
⑧ 服务中心 service center
⑨ 入口广场 entrance square
⑩ 水上平台 water platform

景观设计理念：打造生态友好，具有娱乐、生态保护、雨水滞留和防洪等复合功能的景观。此节点由公共开放空间、中央公园、站前入口广场和生态公园组成，以达到景观设计目标。

Landscape Architecture Design Concept: Develop an ecofriendly, leisurely landscape with functions such as recreation, ecology protection, stormwater detention, flooding prevention. The landscape comprises with general public open space, central park, station entrance square and ecological park to approach landscape architecture design aims.

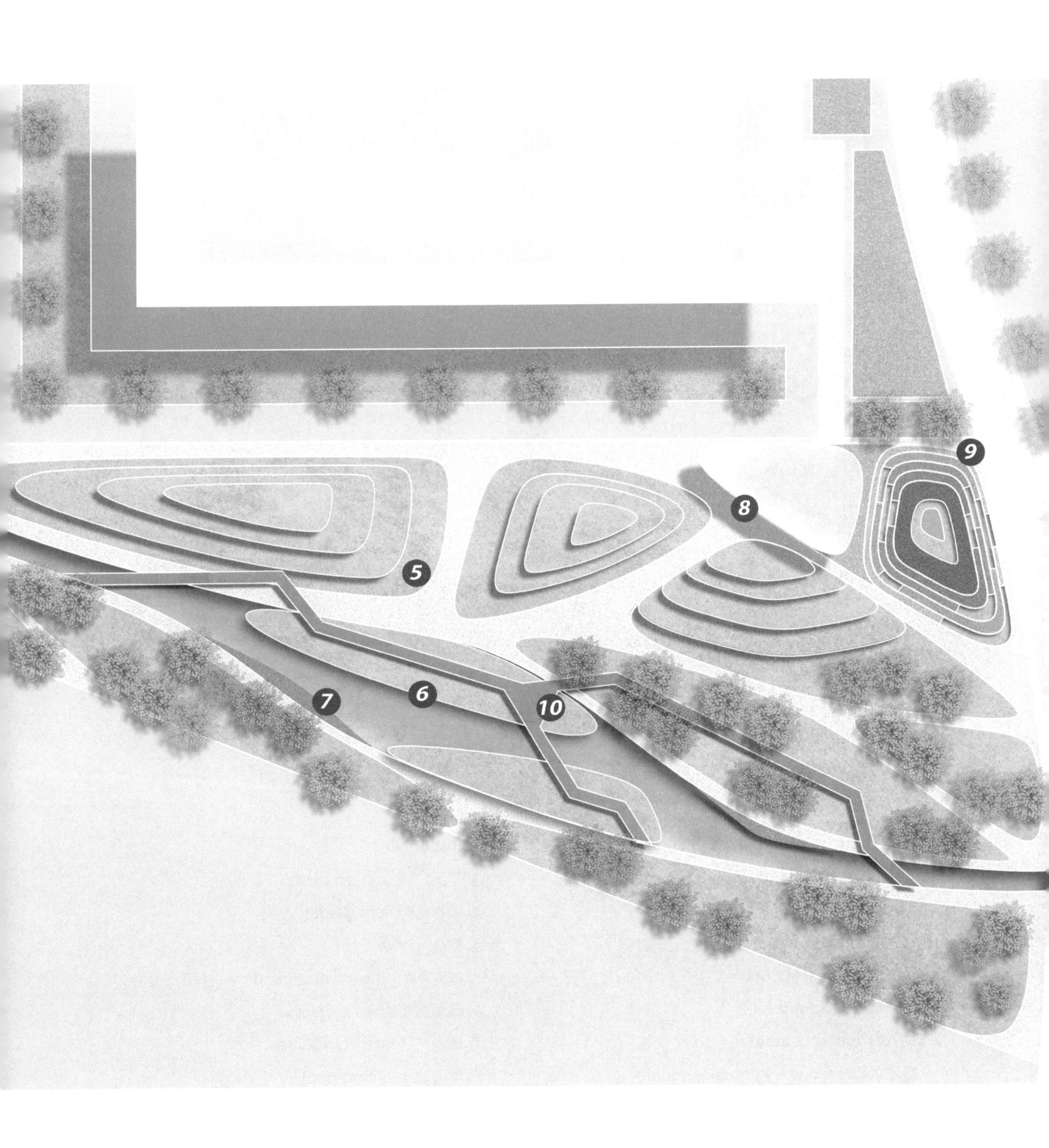
5
8
9
6
7
10

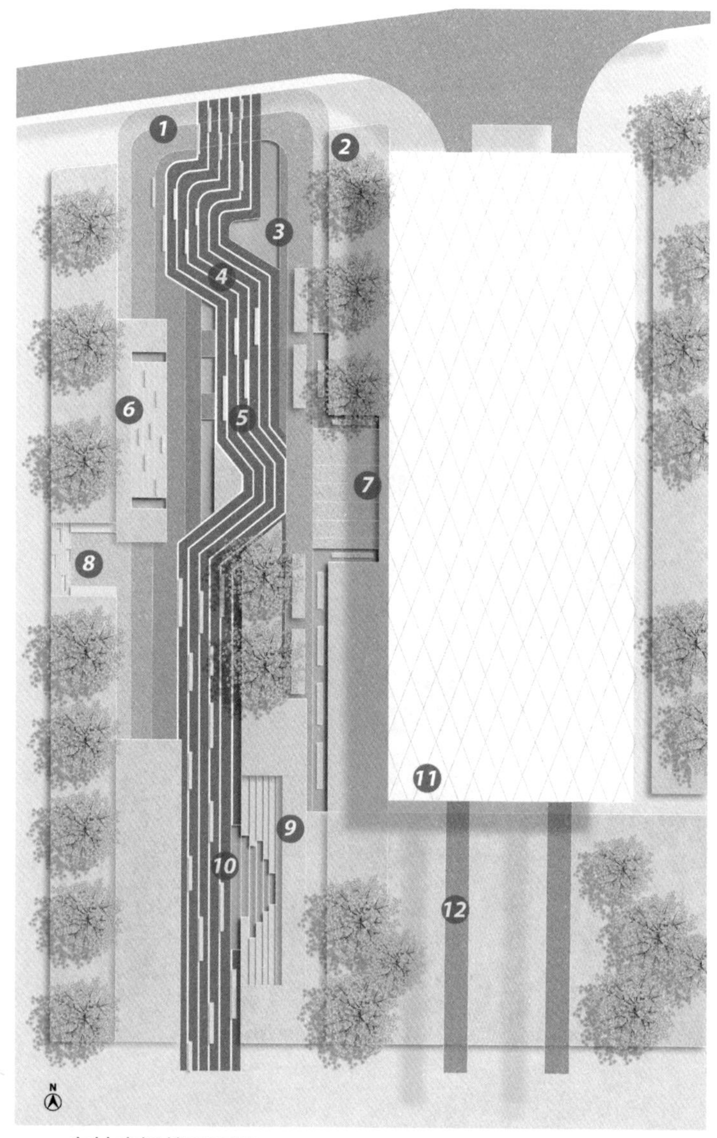

车站广场总平面图
Station Square Master Plan

图例
Legend

① 入口广场 entrance square

② 植被带 vegetation belt

③ 池塘 pond

④ 硬质路面 hard pavement

⑤ 小植物带 small plant belt

⑥ 休憩广场 rest square

⑦ 车站入口 entrance station

⑧ 广场次入口 secondary entrance square

⑨ 楼梯休息区 stairs rest area

⑩ 瀑布 cascading waterfall

⑪ 火车站 rail station

⑫ 铁路 railway

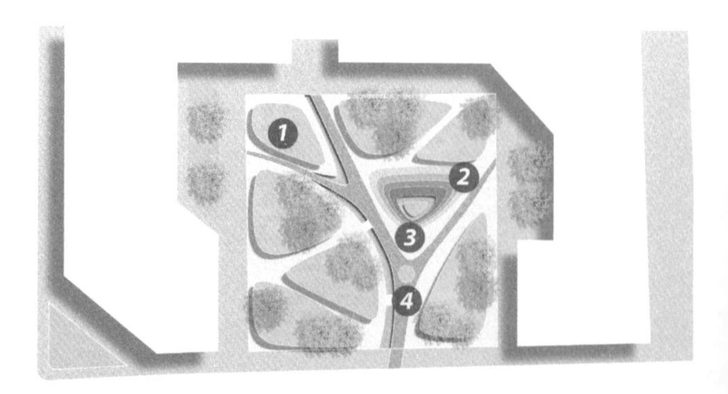

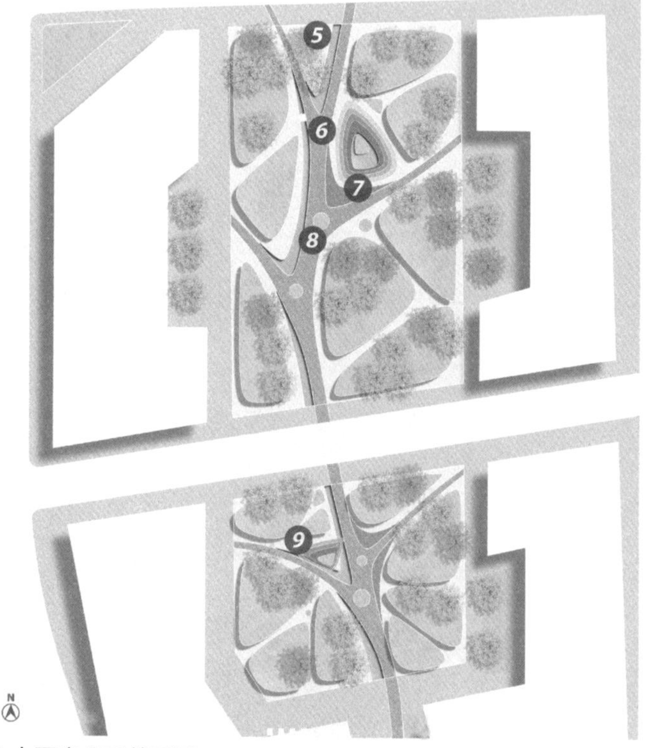

中央医疗公园总平面
Central Healing Park Master Plan

图例
Legend

① 小山坡 hill

② 下沉广场 sunken plaza

③ 雨水花园 rain garden

④ 树丛 grove

⑤ 线性水池 linear water pond

⑥ 硬质路面 hard surfacing

⑦ 活动广场 activity square

⑧ 种植池 planting pool

⑨ 滨水广场 waterfront square

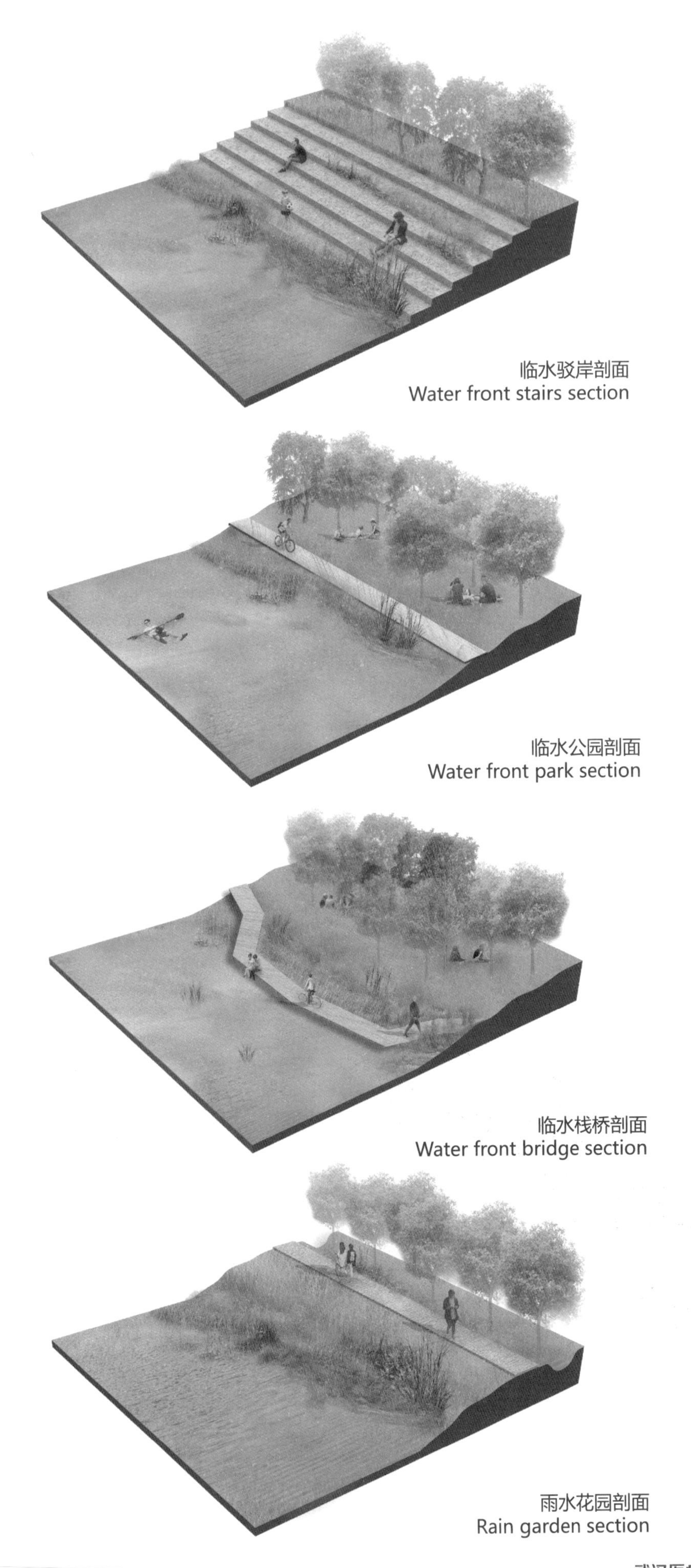

临水驳岸剖面
Water front stairs section

临水公园剖面
Water front park section

临水栈桥剖面
Water front bridge section

雨水花园剖面
Rain garden section

生态公园鸟瞰图
Aerial view of ecological park

车站广场鸟瞰图
Aerial view of station square

中心公园鸟瞰图
aerial view of central park

生态公园效果图
Perspective view of ecological park

车站广场效果图
Perspective view of station square

中央公园效果图
Perspective view of central park

康复医疗区

The Healing Medical District

Gabriel Jenkins

该项目旨在开发和完善一个新型医疗区域，其中包括一些新建的医疗设施：一个新的诊所、一个新开发的医学大学校园和一个康复城市公园。这一设计将促进未来生态健康，并为其提供实质性的保护，以满足武汉不断增长的人口对可持续健康基础设施的需求。

The project aims to develop and strengthen a new medical area，which will include additional hospital facilities—a new clinic, a newly developed medical university campus and a rehabilitation city park. This design meets the growing demand of Wuhan's growing population for sustainable health infrastructure. It will promote an ecologically healthy future and provide substantial care for it.

设计策略 Design Strategy

设计程序
PROGRAM

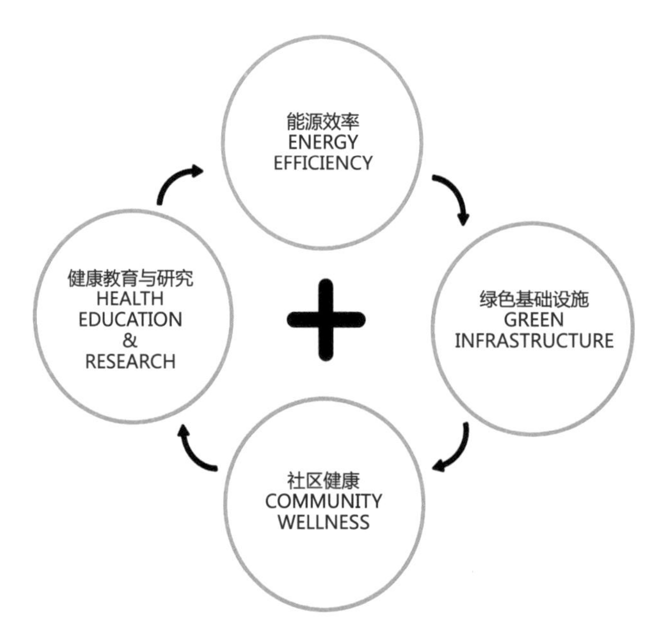

设计策略
DESIGN STRATEGY

健康教育与研究
HEALTH EDUCATION & RESEARCH

连接不同空间，并根据不同的噪声程度、交通状况和主题，利用高程变化创造出不同的区域，为患者及游客提供一个私密空间以体验景观。
Interconnect different spaces and use elevation changes to create areas that vary from noise, traffic, and theme to provide patents along with visitors a private space to experience the landscape.

社区健康
COMMUNITY WELLNESS

构建良好的社交体系以及健康网络，满足病人、家庭以及员工的需要。
Build social and health networks to meet the needs of patients, families and employees.

能源效率
ENERGY EFFICIENCY

构建高效利用资源的景观还有其他益处，比如更低的维护成本，更少的用水量，更安静的场地和更清洁的空气。有效地实施这一点可以极大地增加城市和社区的价值。
Energy efficient landscaping has additional benefits such as lower maintenance costs, a reduction in water use, a quieter site and cleaner air. Effectively implementing this can significantly add more value to the city and community.

绿色基础设施
GREEN INFRASTRUCTURE

绿色基础设施包括植物，如原生植物、药用植物和观赏植物。生态水池通过雨水收集各种水体也能恰当组织场地内的水流，有利于疗愈计划的实施。
Green infrastructure includes plants, such as native plants, medicinal plants and ornamental plants. Ecological pools which collect various water bodies through rainwater, can also properly organize the water flow within the site. It is beneficial to the implementation of the healing plan.

使用人群
USER GROUP

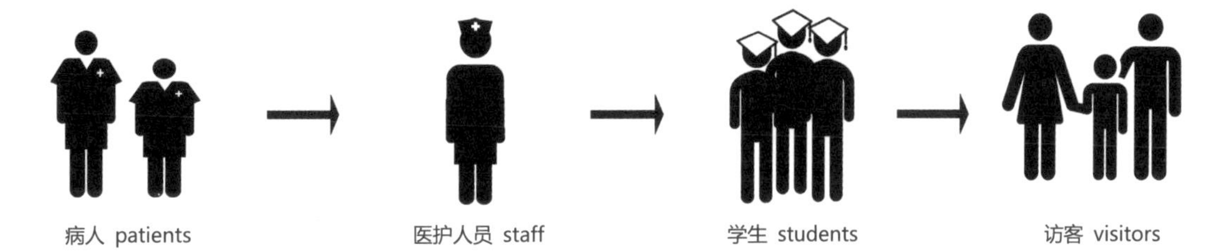

病人 patients　　医护人员 staff　　学生 students　　访客 visitors

设计要点
DESIGN FOCUS

公共和私人绿色空间
public & private
green space

清洁能源
renewable energy

多式联运
multi-modal
transportation

环境可持续
environmental
sustainability

恢复疗愈
HEALING RESTORED

疗愈设计
therapeutic
design

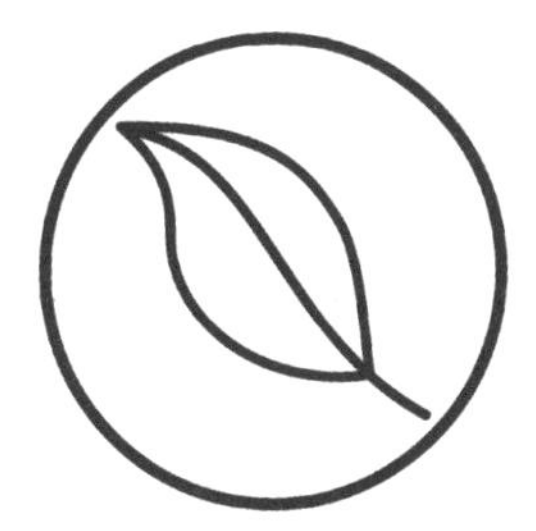

绿色基础设施设计
green infrastructure
design

能源效率设计
energy efficiency
design

种植策略
PLANTING STRATEGY

槭树 *Acer palmatum*

红枫 *Acer palmatum*

依兰 *Cananga odorata*

茼蒿 *Chrysanthemum*

二月兰 *Orychophragmus violaceus*

樱花 *Prunus serrulata*

竹子 *Phyllostachys edulis*

金银花 *Lonicera japonica*

6
10
8
3
7
4
11
1
2

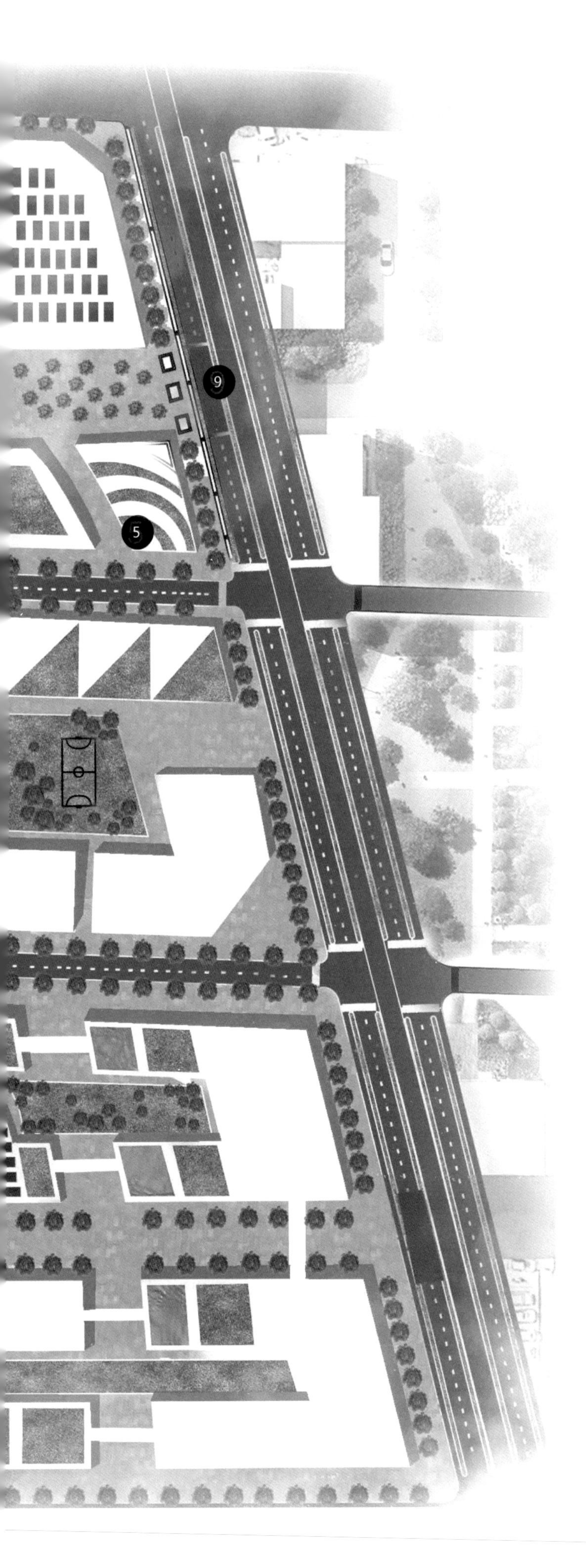

① 改造后湿地/池塘 renovated wetland/pond
② 草地 grassland
③ 林地 woodlands
④ 树林空间 forest space
⑤ 草药咖啡馆 herbal cafe
⑥ 康养花园 courtyard healing garden
⑦ 静思园 meditation garden
⑧ 药圃 herbal garden
⑨ 轻轨站 light rail stop
⑩ 雕塑花园 sculpture garden
⑪ 儿童活动区 children playground
⑫ 户外运动 outdoor athletic fields

专项设计 Design of Special Topic

水系统
Water System

水体 water bodies
径流 runoff

建筑
Buildings

商业楼房 retail buildings
校园公寓 campus residences
校园医疗建筑 campus medical buildings

植被
Vegetation

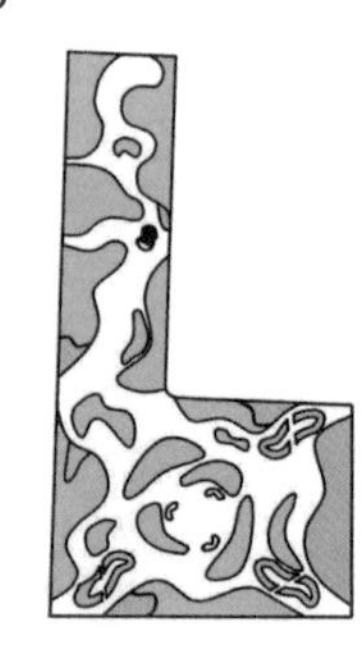

绿色空间 green space
硬质空间 hard space

娱乐
Recreation

开敞活动空间 open recreation space
校园建筑 campus building
私密绿色空间 private green space

交通
Circulation

道路 roads
轻轨线 light rail path

入口
Entrance

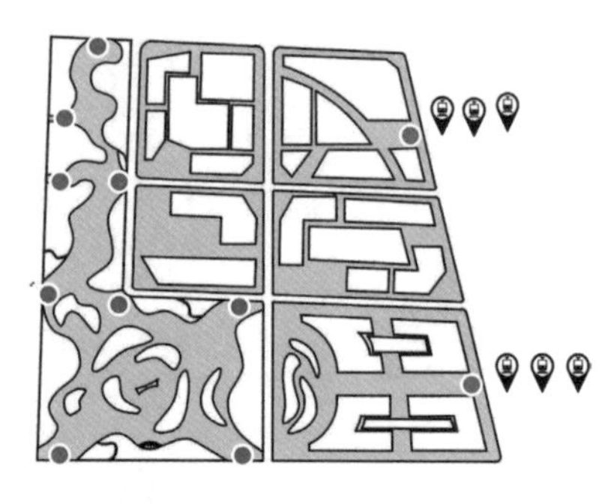

硬质场地 pedestrian hard scape
轻轨站 light rail stop
入口 entrance

剖面图
Section

详细设计 Detailed Design

太阳能设施为轻轨站提供能源
Light rail stop operated and powered by solar energy infrastructure

在城市公园的中心构建人工湿地
Engineered wetland in the center of the urban park

草药咖啡馆作为医疗限制街区的交流点
Herbal cafe that serves as social hub for the medical district strict area

口袋公园促进私密空间内的疗愈行为
Courtyard garden that facilitates the private space healing

第三部分　埃及学生作品

Works from Egypt

长江畔的新兴弹性城市

——应对流行疾病的弹性风景园林与建筑

A Young Resilient City on the Yangtze River

—Pandemic Resilient Landscape, and Architectural Buildings

本次设计初始即对流行疾病的起因及其在人群间传播如此迅速的原因进行了研究，以便为人们提供一个接触较少，但生产力更高的安全环境。这一想法随后被落实为具体的城市项目和建筑项目。城市项目包括不同尺度空间的流行病应对弹性方案，大到城市尺度，小到建筑间的城市空间。建筑项目则包括适用于不同建筑类型的建筑设计解决方案。

The design process started off by researching the cause of these pandemics and the reason why its spreads this fast between people to provide a safe environment for the people with less interaction but more productivity. This idea was then transformed into specific architectural and urban projects. The urban projects include scenarios of resilience to pandemics on different scales starting from the city scale till the urban spaces between buildings. The architectural projects include solutions for different building types that are applied on the building's designs.

长江一项目 Youngtze River - The Project

2020 年初，新型肺炎的暴发，整个世界都措手不及。应对此类突发事件时我们的准备是极不充分的。

In early 2020, The outbreak of the novel coronavirus disease took us by surprise and the world wasn't ready for it. We faced the consequences of not being well prepared for such incidents.

因而建筑和城市规划组织现在都在讨论如何在未来构建起流行疾病之下的弹性城市，以及建筑环境可能的变化。WDS 是最早致力于为这一问题提出解决方案的工作坊之一，以武汉市为例，将拟议的策略作为蓝本应用于世界各地。我们从闻名世界的长江水汲取灵感，针对这座长江新城提出策略。它的确是一座坐落于长江边的新兴弹性城市。

So the architecture and urban planning community is now discussing how to have pandemic resilient cities in the future and how the built environment would change after this pandemic and the WDS was among the first studios to focus on proposing a solution for this problem taking Wuhan city as a case study to apply the proposed strategies as a blueprint to be repeated in other places around the world. So inspired from the famous Yangtze River comes our proposal of the Youngtze City. It is a young resilient city by the river.

项目目的 Project Goals

此项目主要目的是通过建筑和城市规划，为城市抵御流行疾病暴发提出一个完整的导则，从而构建起安全且健康的城市。为了实现这个目标，我们需要构建一座在电力、粮食和水这些资源上可以自给自足的城市，以便能够控制其安全范围并限制流行疾病暴发。

——城市层面：自给自足。

——城市规划层面：根据城市应对流行疾病的情况，为城市主要节点、人群及患者流线和社区规划提供战略设想。

——建筑层面：通过建筑设计限制人群接触，并为其提供安全的工作和生活环境，从而避免暴露于危险之中。

The main goal is to provide a safe and healthy city through a complete manifesto for resilience to a outbreak of a new disease in a city through architecture and urban planning. In order to do that we need to have self sufficient cities, where cities could be self sufficient in power, food and water to be able to control its perimeter and control the outbreak of the diseases.

— Cities scale: to be self sufficient.

— Urban planning scale: to provide strategic ideas for main city nodes, circulation of people and patients, planned neighborhoods according to correspondence of the city to deal with new diseases.

— Buildings scale: to have buildings designed to control people's interaction and provide safe environment of working and living their lives without putting their lives in danger.

项目目标 Project Objectives

此项目的主要目标是构建健康、安全、绿色、繁荣和智慧的城市，从而从源头预防疾病暴发，一旦疫情发生人们依然能维持同往常一样的生活。

——健康：通过提供步行街道鼓励人们采取积极的生活方式。

——安全：通过提出此类疫情的弹性导则，避免因人们的新生活方式引起疫情暴发，从而保证安全。

——绿色：在各处构建绿色走廊和开敞的绿色空间。

——繁荣：发展工业区并建立庞大的国际交流区。

——智能城市：保留和再利用现有功能。

The main objective is to have healthy, safe, green, prosperous and smart cities to prevent the outbreak of diseases in the first place or face it in case it happened without having to pause the people's lives.

— Healthy: by providing walk-able streets and encouraging an active life style for the people.

— Safe: by providing this pandemics resilience manifesto that helps keep the safe by preventing the pandemics outbreak because of the new lifestyle of the people.

— Green: by providing green corridors and large green spaces everywhere.

— Prosperous: developing the industrial zones and creating huge international exchange zones.

— Smart Cities: preservation and reuse of the existing functions.

项目价值 Project Values

此项目主要的关注点是，通过构建一个以新的视角看待社会，减少此类疾病暴发的可能性，从而预防新疾病暴发。因此，项目旨在创建一个在各个方面兼具安全和新视角的城市，包括从整个城市的绿色生态走廊，到新提出的具备不同功能的样地，如工业园区、安全步行区、奥林匹克城市和娱乐文化区。这些既是人与人之间接触最频繁的区域，也应作为疫情时的医疗中心，并考虑到应对疫情的安全措施。

因此，本项目的主要价值是构建一个能够抵御疫病且适用于世界各地的城市蓝本。

The main concern is to prevent the outbreak of new diseases by creating a society that lives with a new perspective towards life that decreases the possibility of such outbreaks. Thus creating a city with safe and new perspective in every aspect starting from the green ecological corridors throughout the city to the new proposed prototypes for different functions such as industrial parks, safe walk-able neighborhoods, Olympics city and entertainment cultural zones were the most interaction between people occurs and medical hub where all safety measures of dealing with a new pandemic is taken into consideration.

The main value is to have a prototype of cities resilient to pandemics that is applicable everywhere around the world.

项目愿景 Project Brief

该项目是中国湖北省武汉市新城市愿景（新武汉）的一部分。场地为一块面积为 15000 费丹（埃及面积单位 1 费丹 ≈ 4200m^2）的土地。从城市战略规划，到按照新提出标准设计特定功能的微观尺度，该区域都将作为疫情下新型弹性城市的样本成为研究对象。

该项目旨在为城市规划、城市设计和建筑设计提出新的标准，以便为人们提供一个以新视角看待生活的安全环境，从而避免流行疾病暴发，并在流行疾病来临之际帮助人们应对。

This project is a part of the new city vision (New Wuhan) located at Wuhan city, Hubei province, China. It is a land of area 15000 feddan, to be studied as a prototype for new pandemic resilience cities starting from strategic plan for this area of the city till the micro scale of designing certain functions according to the new proposed standards.

This project is all about proposing new standards for city planning, urban design and architecture design to help provide safe environments with new lifestyle perspective for the people to prevent the outbreak of new diseases and to help face them in case of its occurrence.

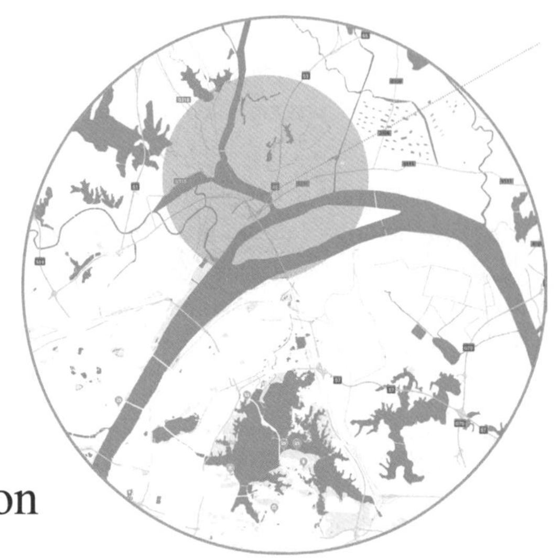

长江新城愿景 Youngtze New City Vision

愿景说明
Vision Statement

打造一个通过构建可步行和骑行的社区来鼓励健康生活方式，从而抵御自然及人为灾害的可持续城市。

Build a self sustainable city of villages that encourages a healthy life style through walkable, recyclable neighborhoods resilient to natural and man made disasters.

任务
Mission

这是由一些“15 分钟”小社区组成的城市，通过步行道和骑行道将街道还给人们，并将繁忙的交通限制在次要道路和主要干道上。

It’s a city consisted of smaller, 15-mins neighborhoods that gives the streets back to the people through walkable, cylable urban streets and restricting high traffic movement to secondary and main roads.

目标 Objectives

流行疾病下的弹性 Pandemics Resilience

通过一些原则从而抵御流行疾病的自给自足型城市

Self Sufficient cities that are resilient to pandemics through some principles

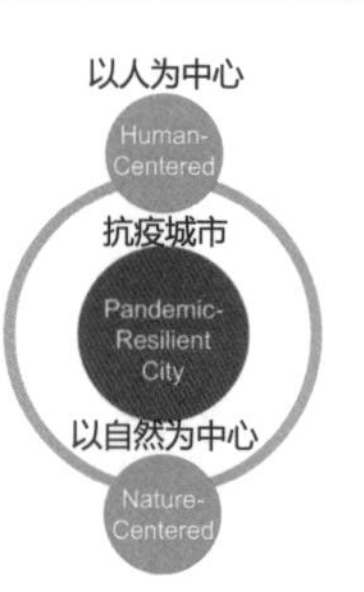

流行疾病下的弹性 Pandemics Resilience

通过一些原则构建自给自足型城市

Self Sufficient cities through some principles

设计策略 Design Strategy

本项目为面积约为12000费丹的整个长江新城制定了规划策略。通过基址分析，我们制定了一些策略：利用绿色元素构建绿色廊道和绿带，通过商业廊道改善商业生活，以文化廊道连接新旧武汉，将城市划分为多个区域保证其在食物、水和电力方面的自足能力和抵御流行疾病的能力，构建利用地铁连接城市所有医疗设施和居民的医疗廊道，通过研发、创新和商业站点来改善贸易循环及经济状况。

This is the strategic plan of the whole Youngtze River New Town, a city of area around 12000 feddan. Site analysis was applied followed by the strategic planning process so we reached several strategies such as green elements as green corridor and green belt, commercial corridor to enhance the commercial life, cultural corridor that connects old Wuhan with New Wuhan, dividing the city to districts that could be self sufficient in food, water and power and resilient to pandemics as well, medical corridor that connects all medical facilities of the city through a monorail as well as other residential, R&D and innovation and business nodes to enhance the trading cycle and the economy.

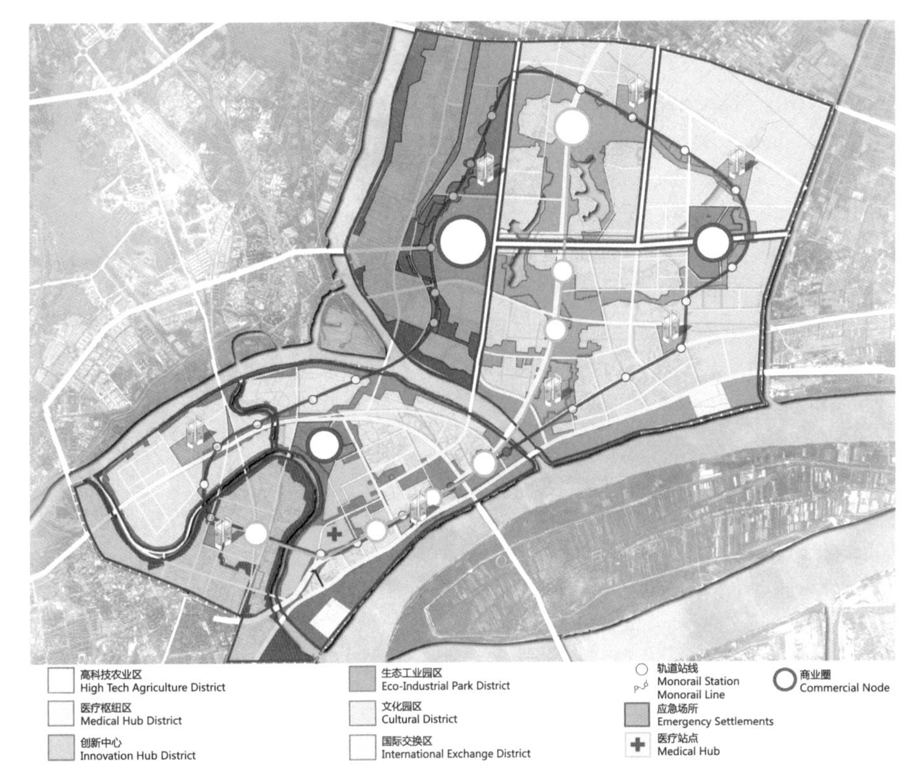

策略应用 Strategies Applied

绿色廊道 Green Corridor

绿带 Green Belt

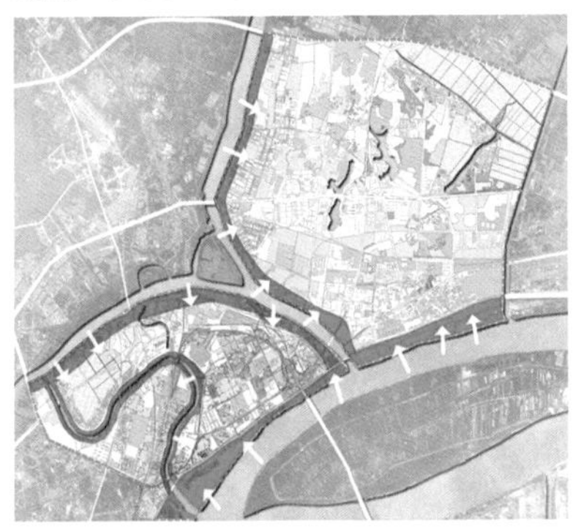

文化廊道 Cultural Corridor

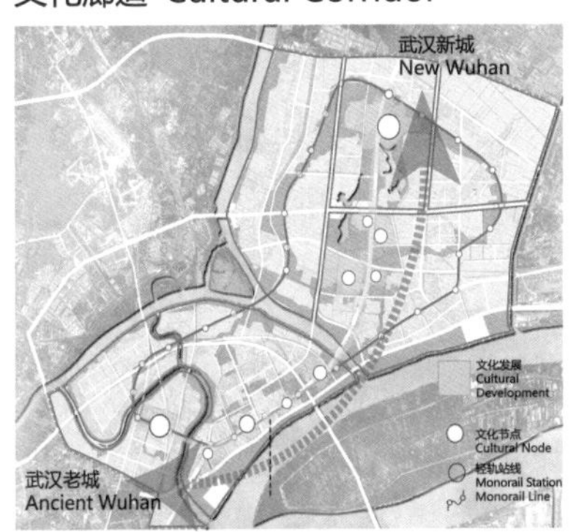

医疗廊道 Medical Corridor

区域划分 Districts Division

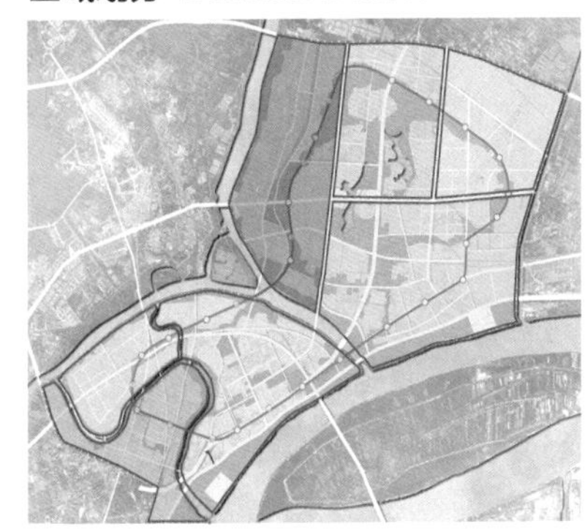

可达性 Accessibility

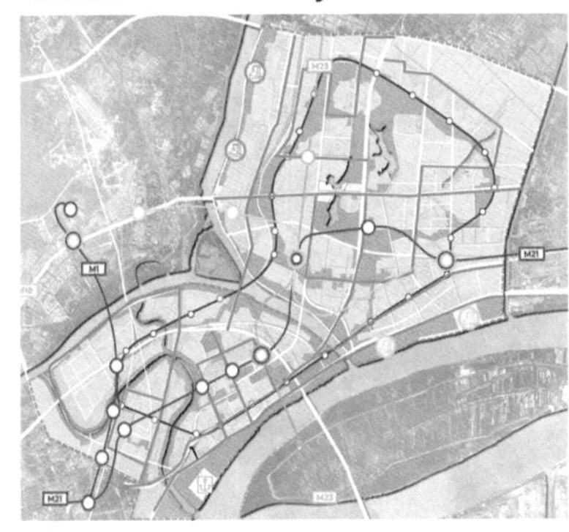

概念设计 Conceptual Design

这是整个城市的概念设计。这七个区都应具备进行所有活动的自足能力，以便在流行疾病来临时能够封闭边界并能自给自足。每个区都具有主导功能和基本功能，包括居民区、教育区、文化区、绿色区、商业区、港口、车站、农业区和医疗区。以国际交流区为例，主要有沿江口岸、住宅区、绿色走廊、博物馆和文化中心等文化用地、医疗单位、应急安置、单轨车站等。

This is the conceptual layout of the whole city. It was designed that each district of the 7 districts is to be self sufficient with all activities so that in case of pandemics it would close on itself and continue production. Each district has a dominant function as well as the essential functions as residential zones, educational zones, cultural zones, green zones, business zones, ports, stations, agricultural zones, medical zones. For Example, the international exchange district has the main port on the river, residential zones, green corridor, cultural lands as museums and cultural centers, medical unit, emergency settlement and monorail stations.

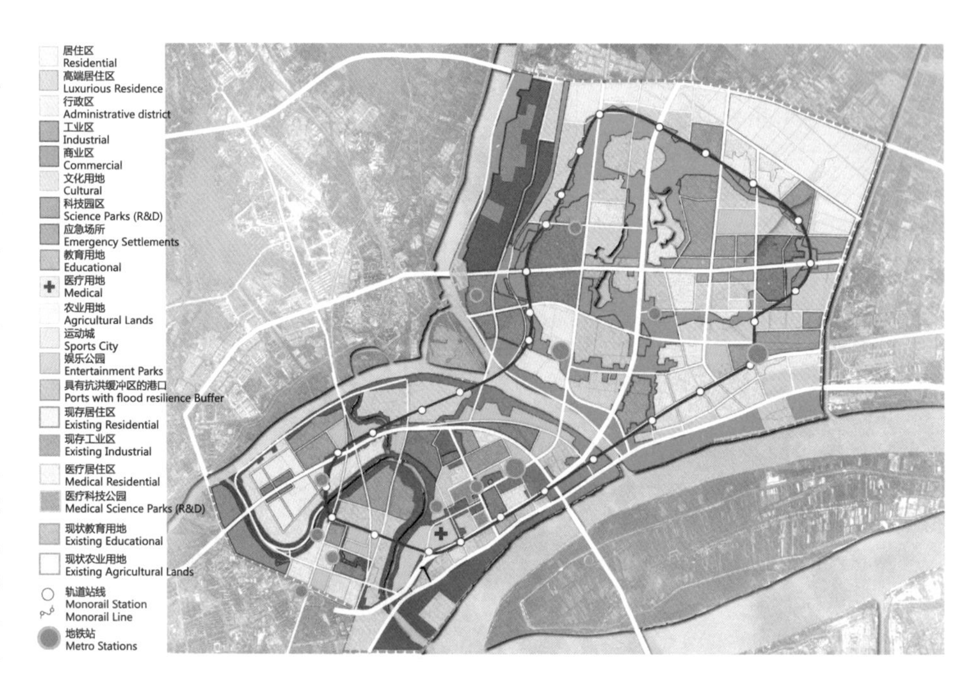

商业区 Commercial Zones

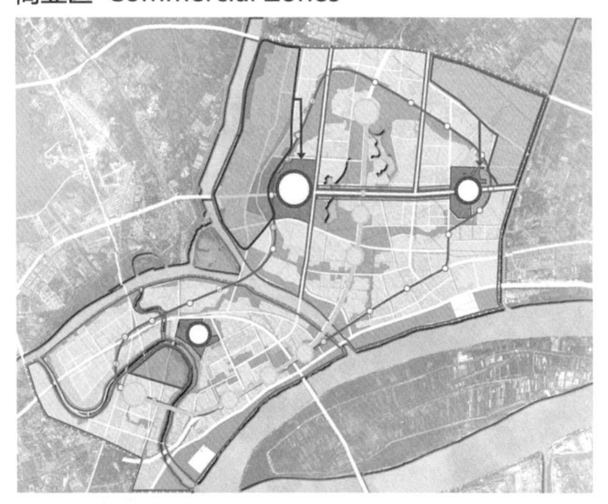

服务区 Service Zones

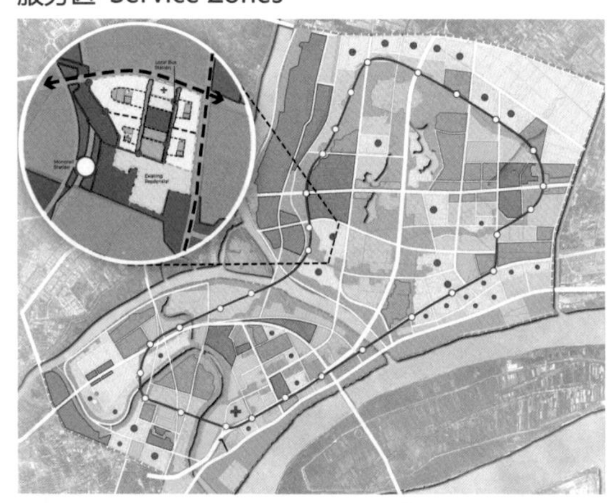

研发区 R&D Zones

医疗区 Medical Zones

绿色廊道活动 Green Corridor Activities

工业区——改造前与改造后 Industrial Zones Before & After

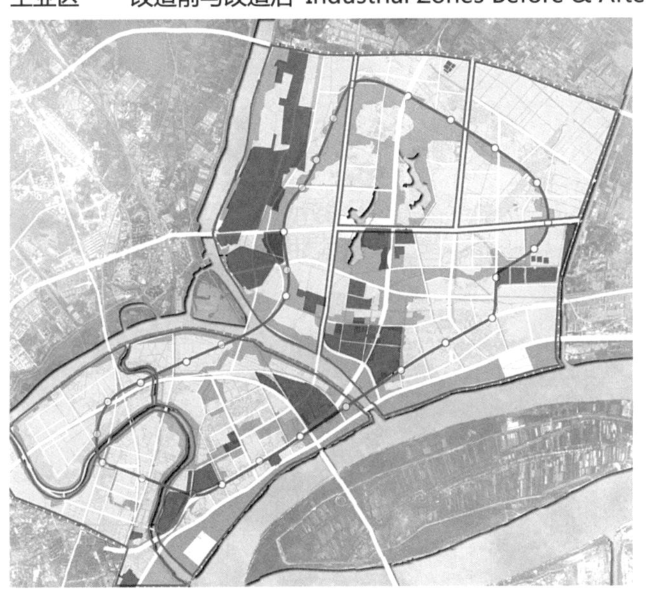

农业区——改造前与改造后 Agricultural Zones Before & After

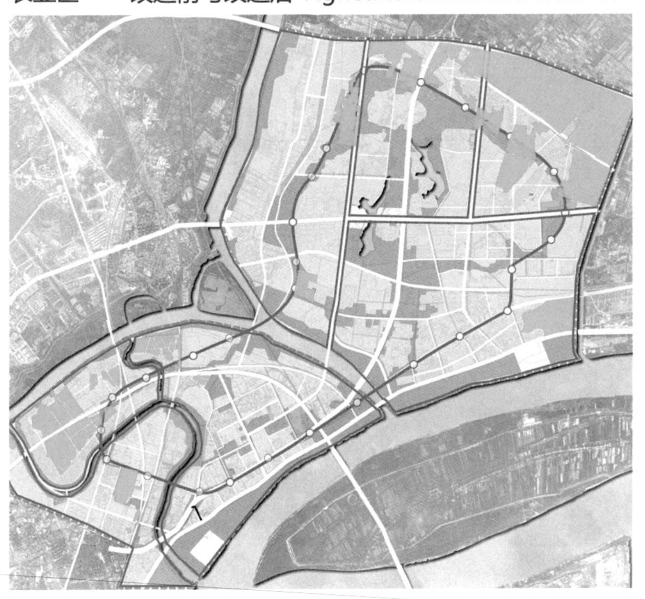

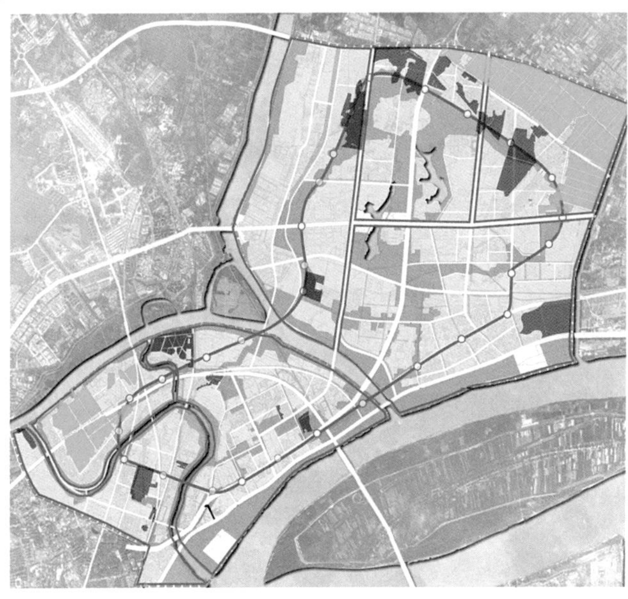

MD
SL
NB
HD
OL
MP

youngtze

弹性问题与主题
——长江畔的新型弹性城市
Problem Resilience & General Theme
—A Young Resilient City by the Yangtze River

项目的该部分旨在假设有与 Covid-19 类似的流行疾病暴发，从正常生活情况（Case 0）到疾病极端暴发（Case C），讨论如何从不同方面处理这些案例，以及从城市环境和建筑方面如何应对此类紧急情况。

整个过程共分为 4 个阶段，分别为 Case 0、Case A、Case B、Case C。

每个案例包含城市策略和建筑策略两部分。从城市尺度到区域尺度，再到建筑物和城市空间的应用策略，直到个人尺度。所有这些都是为了以最少的接触保证人们的安全和生产力以及城市的清洁。

This is the part of the project where it aims to assume an outbreak of a new pandemic as Covid-19 and cases of this outbreak from the normal life case (Case 0) till the extreme outbreak of the disease (Case C). Putting strategies of dealing with these cases in different aspects and how the built environment with its urban side and architecture side will deal with such emergencies.

It was divided into 4 cases from Case 0 to Case A and Case B till Case C.

Each case was divided into urban strategies and architecture strategies. Strategies were applied from the scale of the whole city to the districts to the buildings and urban spaces till the human scale. All is to keep the people safe and productive with least amount of interaction to maintain a clean city.

专项设计 Case 0

这是任何疾病暴发前的正常生活情景。这一情景最为重要，因为新的生活视角从此开始，从而改变人们的生活、工作和生产方式，做好准备，防止流行疾病暴发，并为可能发生的流行疾病做好应对准备。

This is normal life case before any outbreak. This is the most important case because this is where the new perspective of life must be applied so that the life style of the people must change to be ready to work and produce to prevent outbreak of pandemic and to be ready for resilience to these outbreaks if it happened.

不同生活方面的策略地图 Strategies map in different life aspects

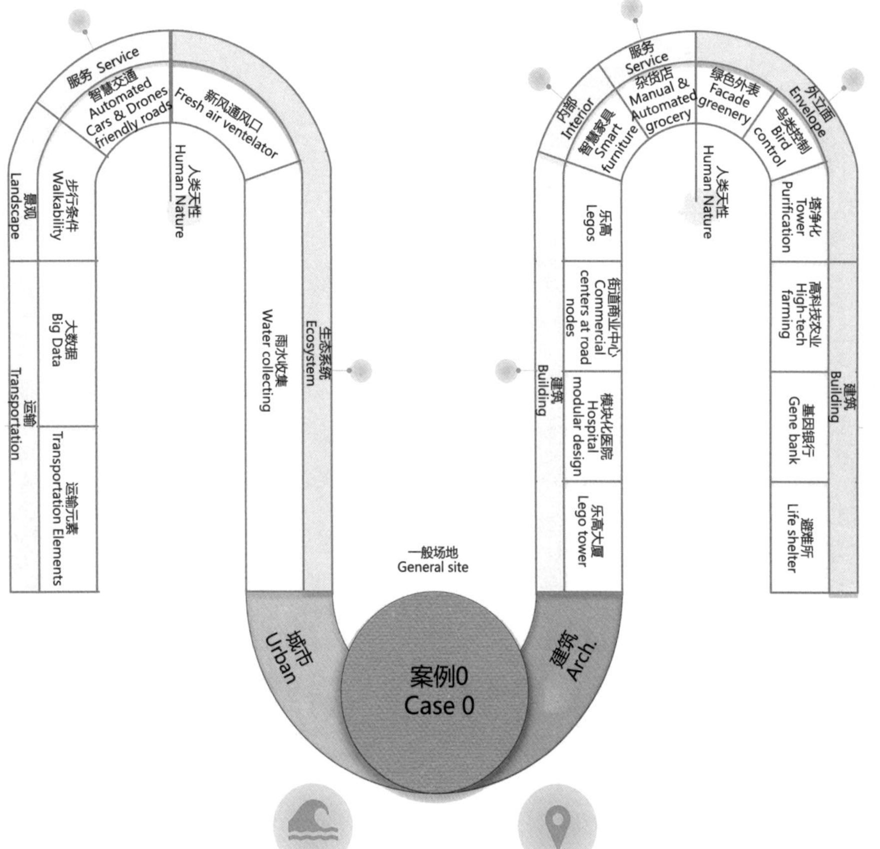

交通方式 Transportation Method

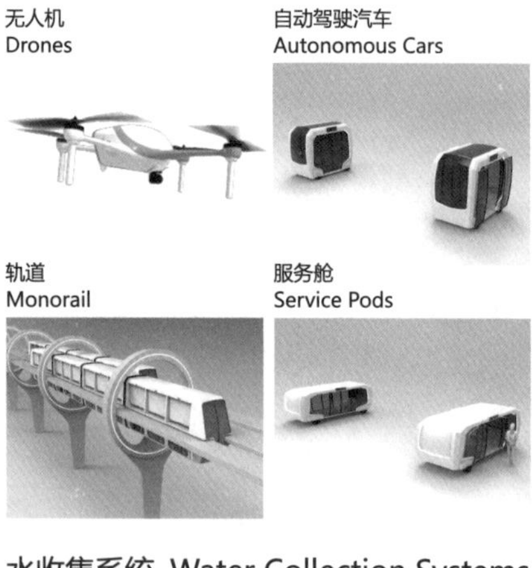

水收集系统 Water Collection Systems

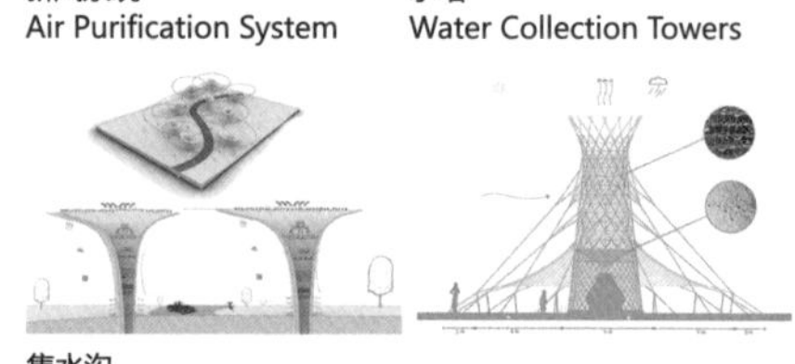

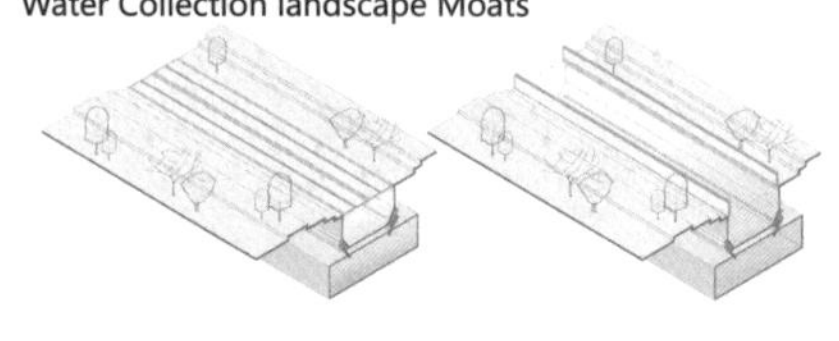

建筑设计 Architecture Design

将如乐高积木般的可移动单元用于景观和建筑。

Legos and adaptable movable units are used in landscape and in buildings.

模块化住房 Legos and Lego Houses

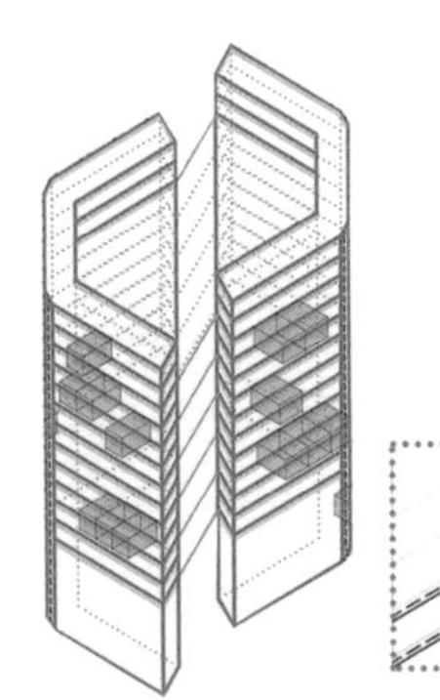

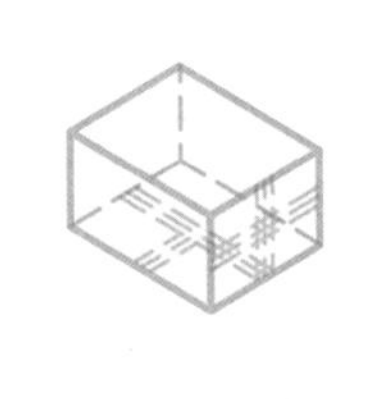

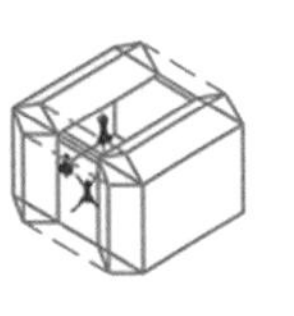

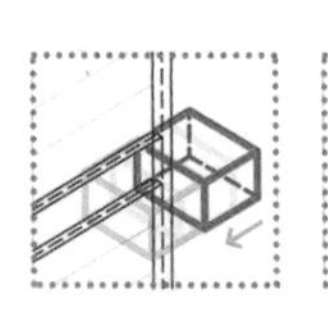

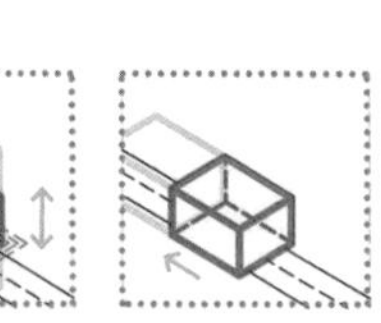

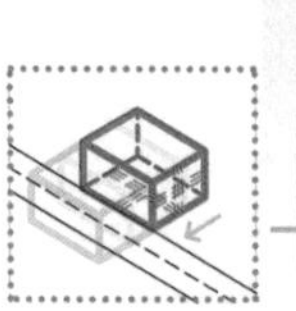

模块化设计 Modular Design

建筑物的模块化设计，便于在紧急情况下进行扩展和使用。

The modular design of buildings could be expanded and used in cases of emergency.

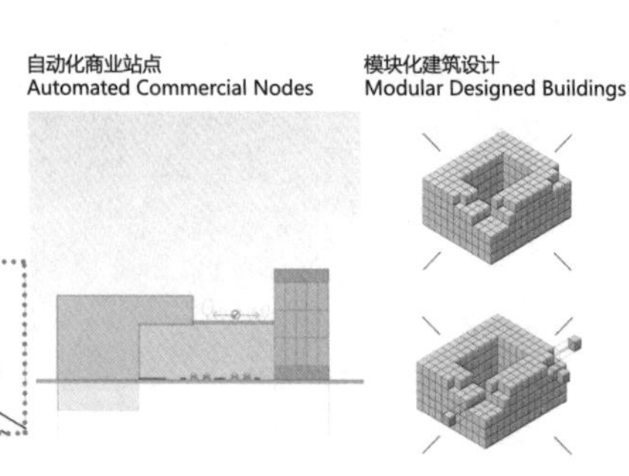

专项设计 Case A

这是抵御流行疾病暴发的第一种情景。在疾病发生且在周边地区泛滥的情况下，整个城市受到了保护且保证了其不被污染。此情景下所有方案都适用于当前及之后的所有阶段，例如 Case A 的任何解决方案对 Case B 和 Case C 都适用，依此类推。

This is the first case scenario of resilience to pandemic outbreak. This is the case where the whole city is protected and clean while the disease is still out there and may be in surrounding areas. All scenarios of cases are applied in the following case, for example, any solutions in case A are applied in both case B and case C and so on.

不同生活方面的策略地图 Strategies map in different life aspects

景观 Landscape
广场封锁 Plaza lockdown
新风系统 Fresh air ventelator & Disinfectant columns
人类天性 Human Nature
生态系统 Ecosystem
水盾 Water shield
运输 Transportation
道路节点 Roads node
储水 Stored water
病毒包围下被保护的地点 Protected site-Surrounding with Viruses
城市 Urban
案例A Case A
建筑 Arch.
建筑 Building
可转换建筑 Convertible buildings
人类天性 Human Nature
可控制的庇护所 Controlled Life shelter
鸟网 Bird Netting
外立面 Envelope
塔净化与消毒 Tower purification & disinfection

空气净化 Air Purification

建筑外立面
Buildings Envelopes
塔净化与消毒
Tower Purification and Desinfectant

城市屏障 Barriers between cities

对抗屏障
Confrontation Barrier

道路节点 Road Nodes

城市周围和主要道路上的道路节点，用于检查进出人员。
Road Nodes around the city and on main roads are to check people moving in & out.

紧急设施 Emergency Settlements

可插入式自由单元结构
Free Structures to plug units in

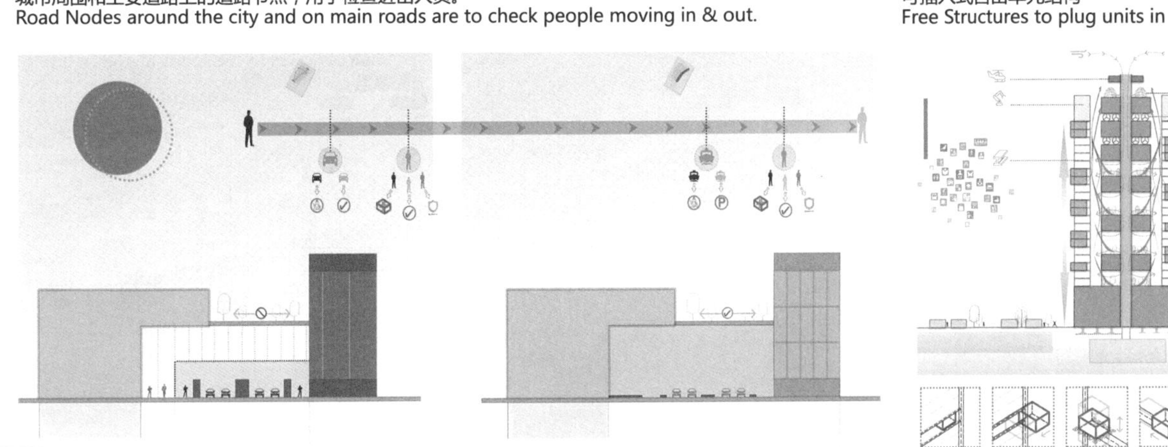

专项设计 Case B

这一情景是在流行疾病暴发更加严重的情况下对其弹性能力的设想。在这种情景下，只有一个感染区，但其他区是安全且未受污染的。这意味着流行疾病开始进入城市，因而每个区域都应采取不同策略予以应对，同时保证区域内的安全和生产。在前面的情景中应用到的所有策略，除深入策略外，在此情景下都有所使用。

This case is the next level of cases of pandemic outbreak and its scenarios for resilience to this pandemic. This is the case where there is an infected district but the rest of districts are safe and clean. This case means the pandemic is starting to get in the city so each city district needs to deal with it and start applying the different strategies to keep the district safe and productive.All strategies applied in previous cases are used in this case in addition to further strategies.

不同生活方面的策略地图 Strategies map in different life aspects

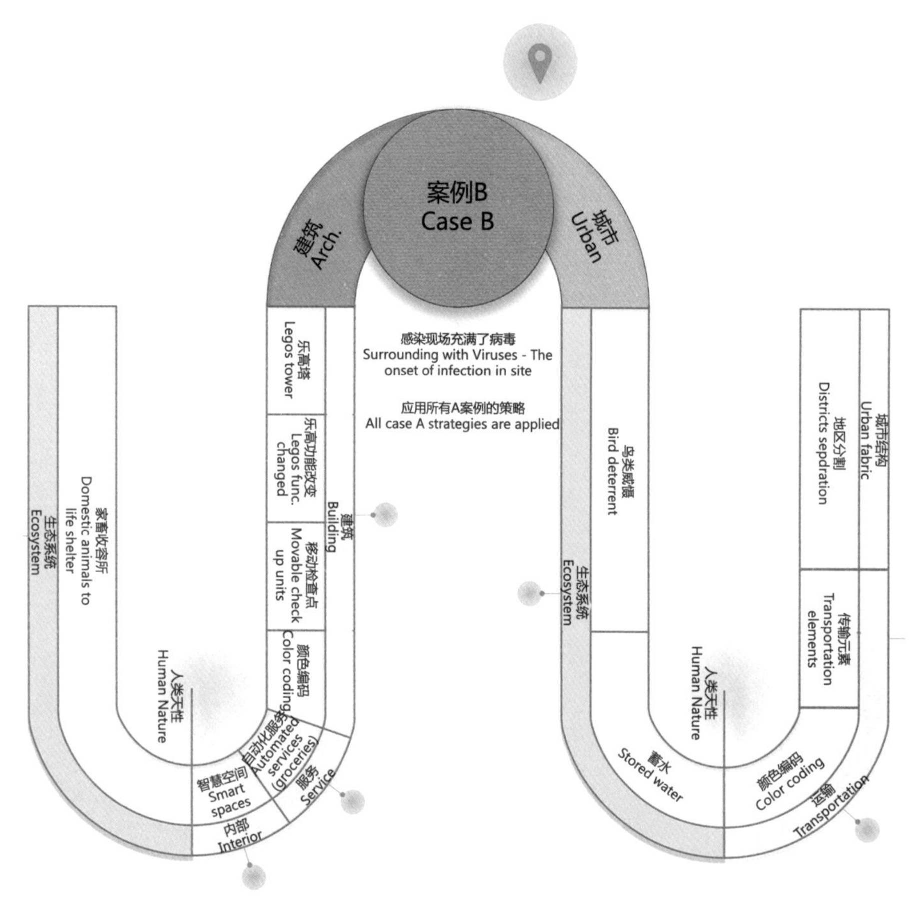

弹性智慧空间 Adaptable Smart Spaces

内部设计
Interior Design

智慧空间 Smart Spaces

夹层作为工作空间或者隔离空间
Mezzanine used as: Working space or Quarantine room

可转换家具
Convertible furniture

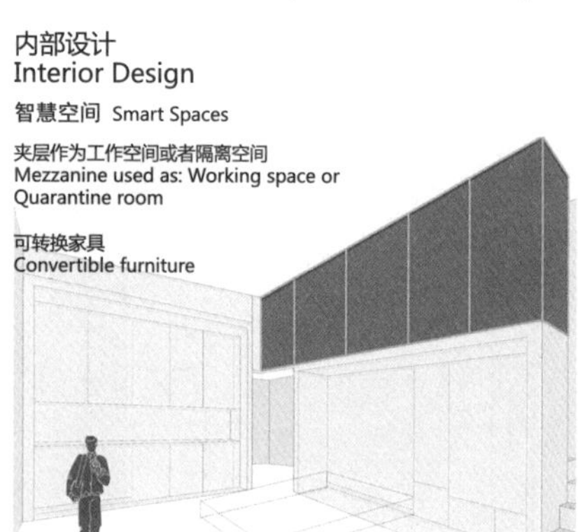

大数据技术 Big Data Technology

大数据用于识别公共场所的感染病例。
Big Data is used to identify the infected cases in public areas.

雨洪韧性和滨水节点 Flood Resilience and Water Nodes

水屏障和滨水节点用于检查从水边的游客。
Water Shield and water nodes are used to inspect visitors from waterside.

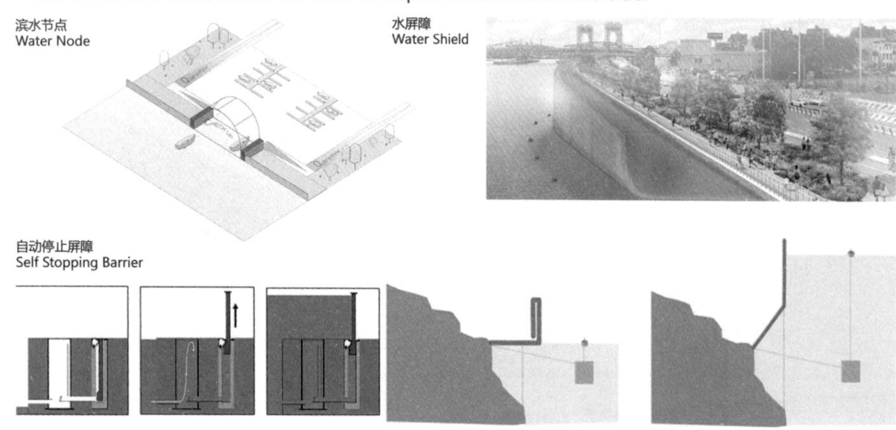

感染者运送 Transporting the Infected

将感染者用电动自动隔离舱送往医疗单位或紧急安置处。
Electric automated pods are used to deliver the infected ones to the medical unit or emergency settlements.

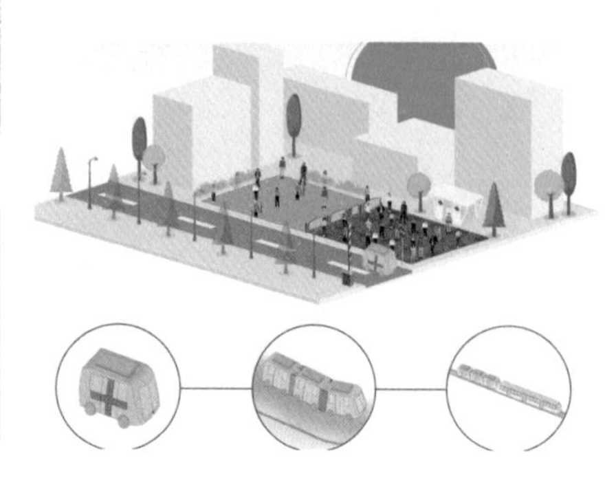

专项设计 Case C

这是流行疾病暴发时最为严重的一种情景。在此情景下，所发生的都是极端状况且逐步失控，人们必须待在家。要制定隔离规则，并将可变公共空间和建筑搭建为临时医院来治疗患者。同之前的情景一样，除了深入策略外，前面所提及的策略也可用于此情景。

This is the most severe case of pandemics outbreak. This is the case where all scenarios are extreme since the situation is going out of control and there is a huge need to keep people in their homes and set quarantine rules and start to use changeable public spaces and building to hospitals to help treat patients. As well as previous cases the previous strategies are used in this case in addition to further strategies.

不同生活方面的策略地图 Strategies map in different life aspects

可折叠检查单元 Foldable Checkup units

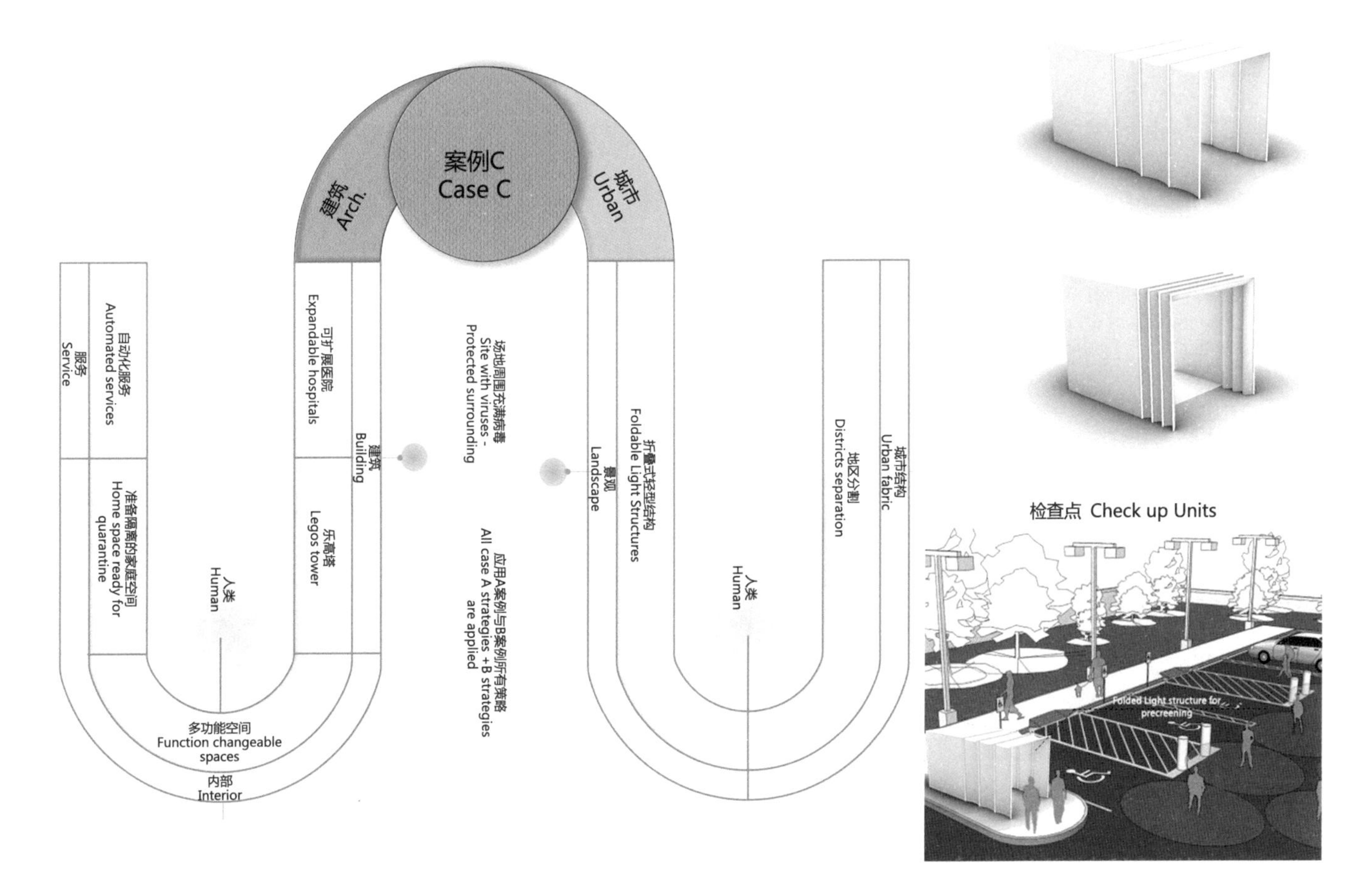

可扩展的建筑设计 Expandable Architecture Designs

通过可扩展的模块化建筑（尤其是医疗类），为病人提供合适数量的空间。
Expandable modular buildings especially medical buildings are to harvest suitable number of spaces for patients.

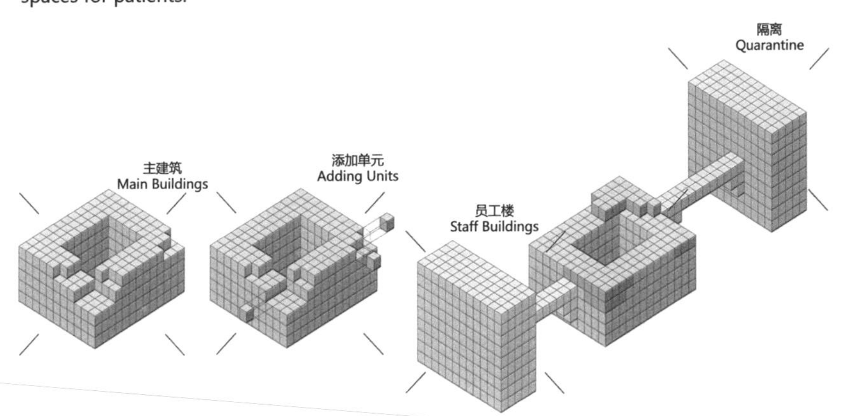

无人机运输 Drones Transportation

采取无人机送货上门，以减少人与人之间的接触。
Drones are used to deliver goods to people in their homes to limit the interaction between people.

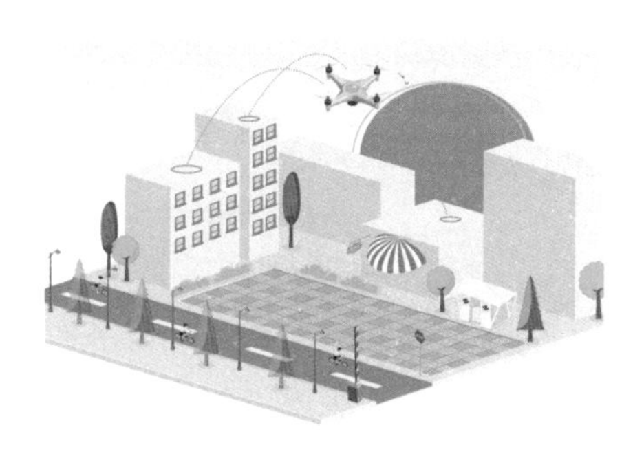

主题与概念 General Themes and Concepts

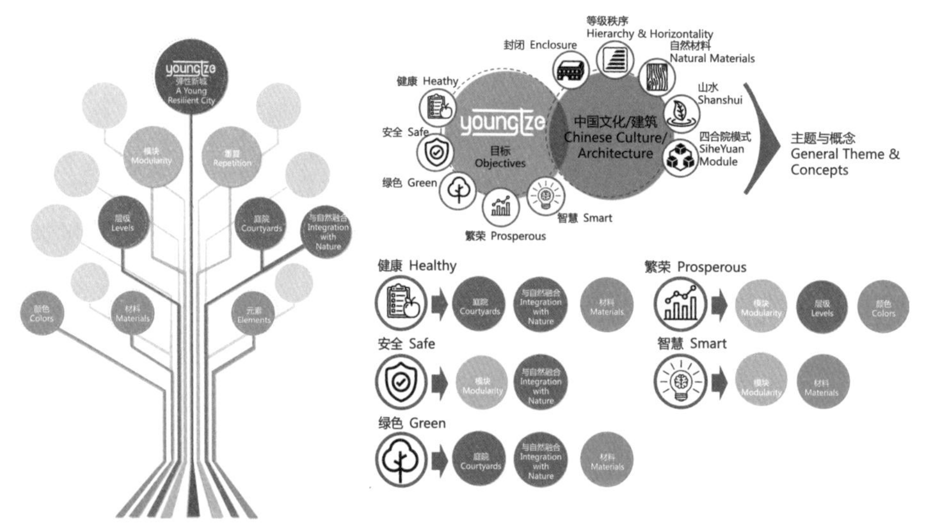

主题受到中国传统建筑理念的启发，旨在以此实现项目目标。

The general theme was developed to achieve the project's objectives in a way that is inspired by the rich traditional Chinese architecture aspects and principles.

这些概念和主题都是从中国建筑中汲取的灵感，包括其模块化的形式、色彩、层次、元素以及文化中蕴含的精致细节。

These are all general concepts and themes inspired from Chinese architecture from modularity and levels to colors and elements with all its fine cultural details.

模块化 Modularity

每个区域都将模块化作为其主要设计概念，但又在各方面采取了各不相同的方法。这一点无论是在建筑的设计和施工上、景观上，抑或在功能的集约化和原型化上，都有所体现。

Each zone used the modularity as its main concept of design but with different methods and in different aspects whether in buildings design and construction or in landscape or in clustering and prototyping the functions and so on.

院落 Courtyards

院落也是受中国建筑启发的概念之一，但在设计中运用了更现代的方式保证持久的自然采光或空气流动，并在户外增加同自然融为一体的社交区域。

Courtyards are also one of the concepts inspired from Chinese architecture but it was used in a more modern way for reasons as enduring natural lighting or air movement or increasing the social areas integrated with outdoor and nature.

层次 Levels

受中国等高线、自然地形层次和建筑中楼梯的启发，每个区域都使用了层次概念，并在不同方面采用了不同的方法，例如用于连接建筑物的坡道和桥梁或是景观元素。

Each zone used the concept of levels inspired from Chinese contour lines, levels in nature and stairs in architecture with different methods in different aspects. Some used this concept in ramps and bridges connecting the buildings others used in landscape elements and so on.

与自然交融 Integration with Nature

这也是在所有区域都有应用的另一概念，增加绿色空间的比例，加强使用者与自然之间的相互作用，并通过可到达的绿色屋顶、绿色桥梁和公园将建筑物与自然融合。

This is also another general concept that got tackled in all zones and it is the concept of increasing the percentage of green spaces and the interaction between the users and nature as well as the integration between buildings and nature through accessible green roofs or green bridges and parks and so on.

色彩 Colors

受中国建筑和文化启发，该项目在建筑立面或景观元素等方面使用红色和黄色。

Red and yellow are colors used in different aspects in the project in some architectural elements in facades or in elements in the landscape and these are colors inspired from the Chinese architecture and culture as well.

材料 Materials

取材于中国传统建筑材料，某些立面和景观元素采用了木头和竹子等材料。

Also materials as wood and bamboo were used in some facades and landscape elements as they were inspired from traditional building materials in Chinese architecture.

详细设计 Selected Zones

该城市被划分为多个区域，分别为文化区、医疗区、社区、工业区、运动区、农业区和创新区。每个区域在水、食物和电力方面都具备足够的自足能力，并且具有与其他区域不同的区域主导功能。我们选择了5个具有不同功能的场地，合理解决建筑和城市设计问题，展示生活和建筑的新视角，最终构建一个可以抵御疫情暴发的城市。

The city was divided into several districts Each district is self sufficient in water, food and power and has a main dominant function that differs than the other districts. The city had cultural zones, medical zones, neighborhoods, industrial zones, sports zones, agriculture zones, and innovation zones. We chose 5 sites with different functions to tackle them architecturally and urban design wise to show the new perspective of life and architecture to form a city that is resilient to pandemics outbreak.

长江新城 面积50平方千米
Youngtze River New City with area 50km²

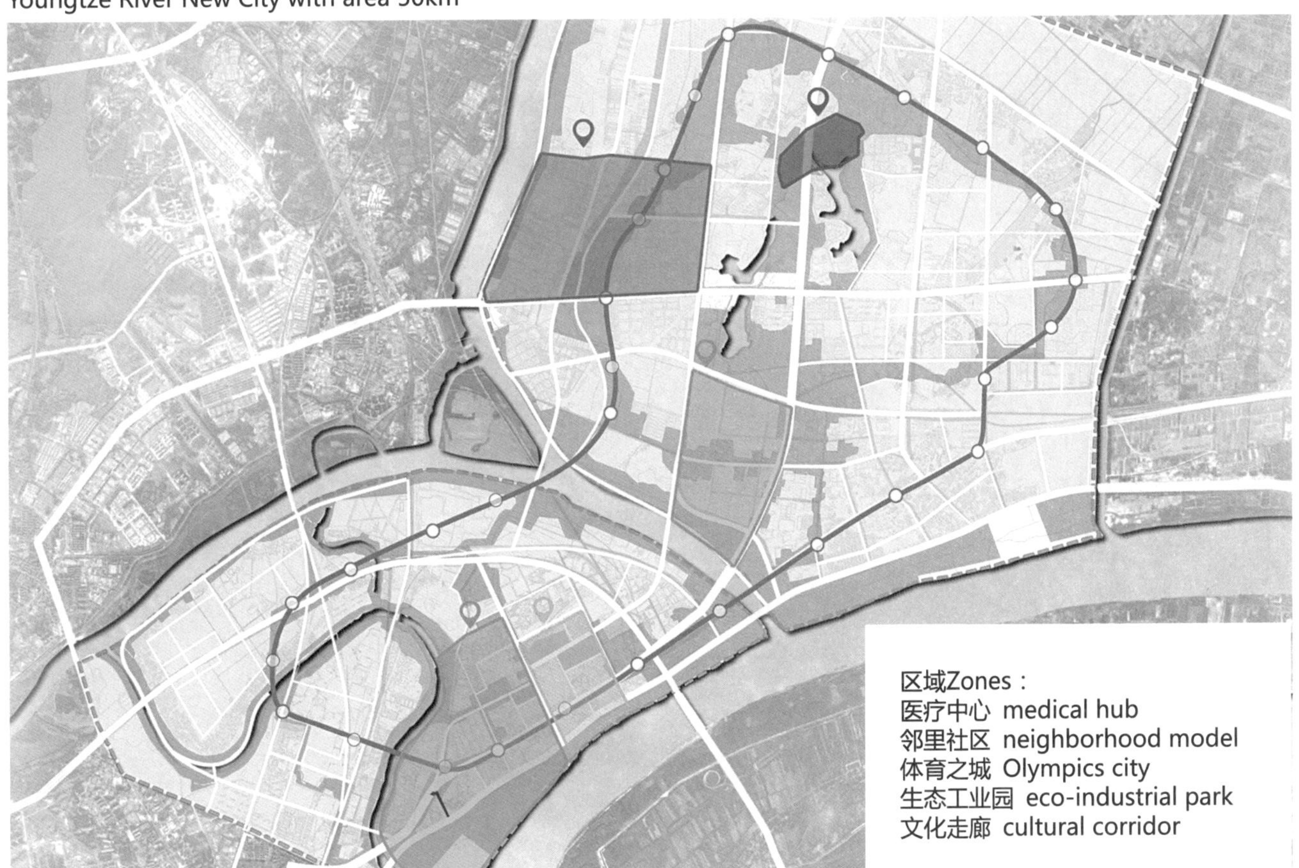

医疗中心

Medical Hub

为了健康，永不停止关怀

A Care that never quits, just for the health of it.

这是一个模块化、灵活、统一的城市，传播高质量的知识型医疗健康生活化服务，并可以应对不同的情况。

具备应急能力的城市包含三个主要区域：

医疗区、创新区和绿色健康廊道。

灵活的模块化单元，可以转换为多种功能并可以在紧急情况下扩展，对城市进行维护。

在人员综合和以环境为中心的护理方面发挥领导作用并取得卓越成就，注重教育、研究、创新和护理服务，营造协作、可持续、跨行业的生态友好环境。

It's a modular, flexible, unitized city to spread a quality knowledgeable-medical healthy life-style services and it can adapt to different cases.

An emergency-prepared city consist of three main zones:

medical, innovation and awareness shaped by a green and a health corridor.

A flexibility modular unit able to convert into multiple function and to extend to maintain in emergencies.

A leadership and excellence in comprehensive people and environment-centered care through education, research, innovation and care provision in a collaborative, sustainable, inter-professional eco-friendly environment.

设计分析 Design Analysis

分区 Zoning

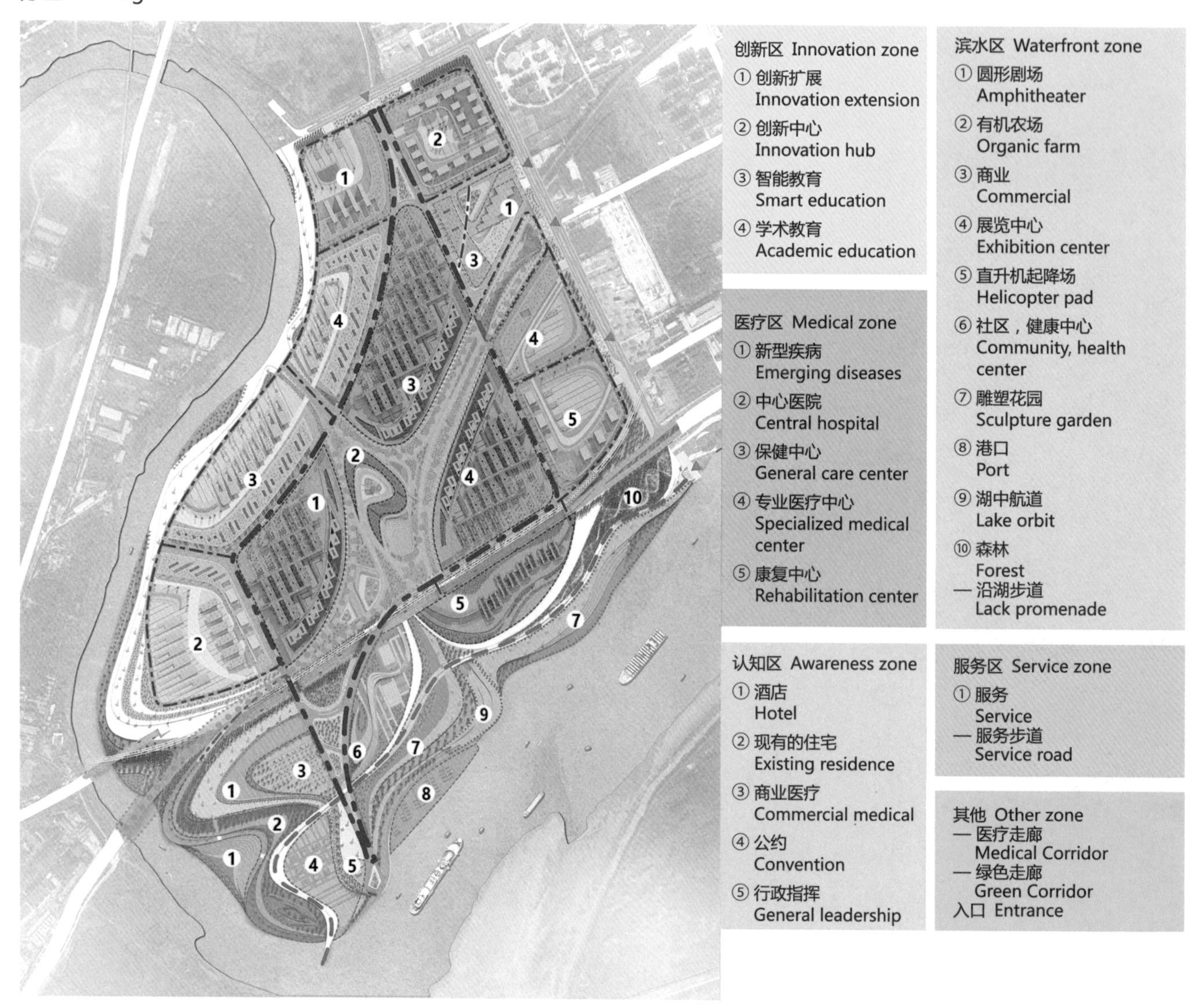

该项目功能分区的思想核心是为城市建立一个医疗中心，配备必要的医疗设施。该医疗中心是可持续、健康且安全的，并且具备对新型疾病和新标准的适应力。

The main zoning idea behind this project is having a medical heart for the city with all medical facilities needed. It will be the most sustainable, healthy and safe medical hub as well as its resilience to emerging diseases and new standards.

可达性 Accessibility

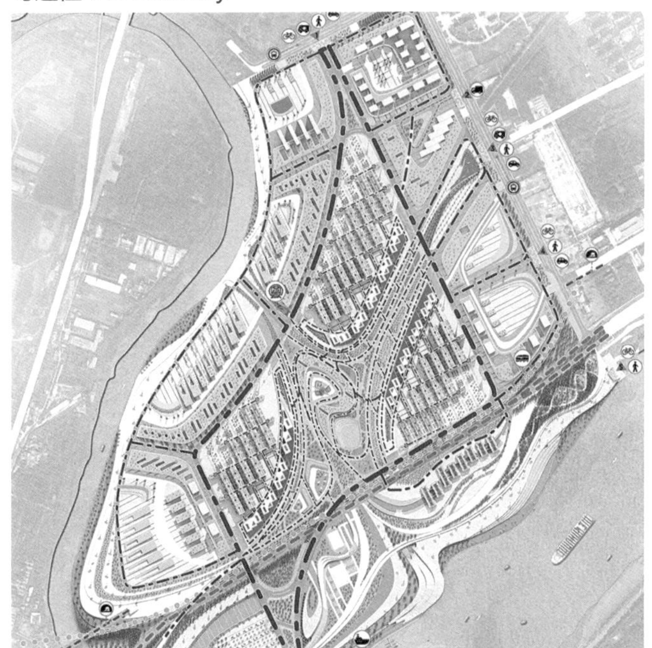

模块化 Legos Automation

鸟瞰图 Aerial View

绿廊 Greenery

流行病扩散 Pandemic Expansion

鸟瞰图 Aerial View

疫病弹性案例 Pandemic Resilience Cases

专项设计 Case 0

医疗中心是根据可能发生的最坏情况设计的。Medical hub is designed based on worst case scenario that could happen.

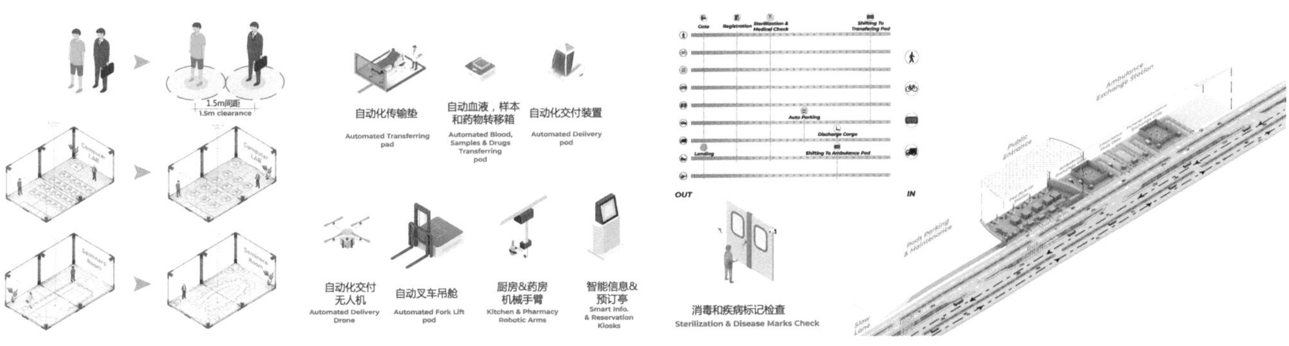

专项设计 Case A,B,C

流行疾病开始暴发时，医疗中心开始应对。This is when the outbreak starts and the medical hub starts to deal with it.

转为医疗单位 Conversion to medical units

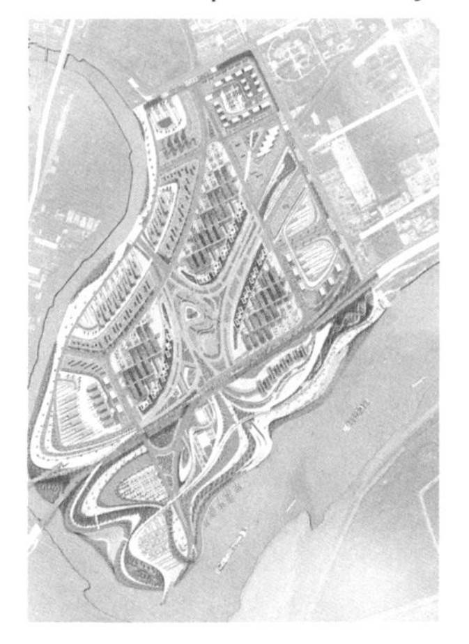

扩展模块建筑 Expansion legos houses

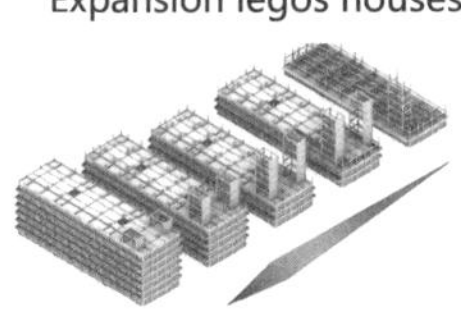

隔离 Barriers

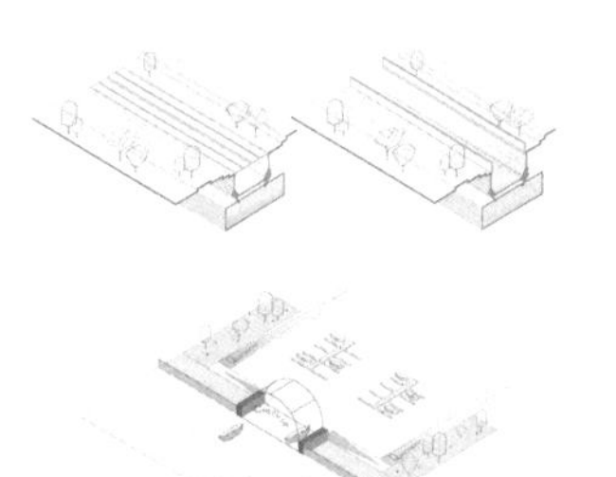

移动式检验装置 Movable check up units

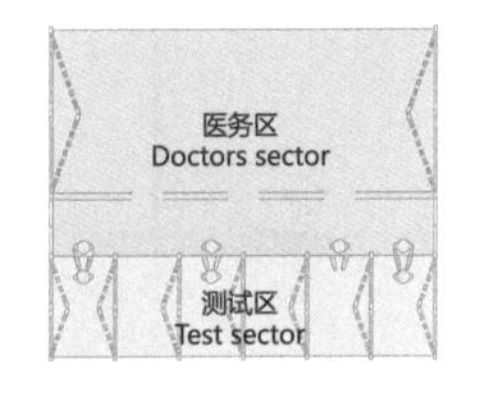

布局转变 Expansion in Layout

渲染效果 Rendered Shot of Expanded Buildings

扩展前鸟瞰图 Aerial View before Expansion

扩展后鸟瞰图 Aerial View after Expansion

滨水设计 Waterfront Studies

景观设施 Landscape Elements

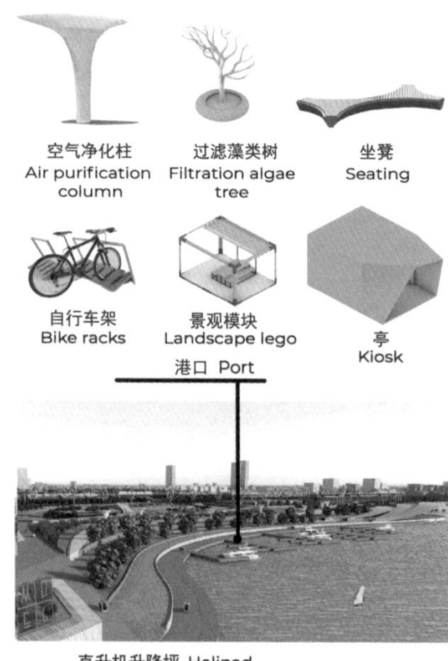

滨水分区 Waterfront Zoning

森林 Forest

运动健康中心 Sports & Wellness Center

可达性设计 Accessibility Studies

道路交通 Rails of Units and Pods Routes

入口区 Entrance Zoning

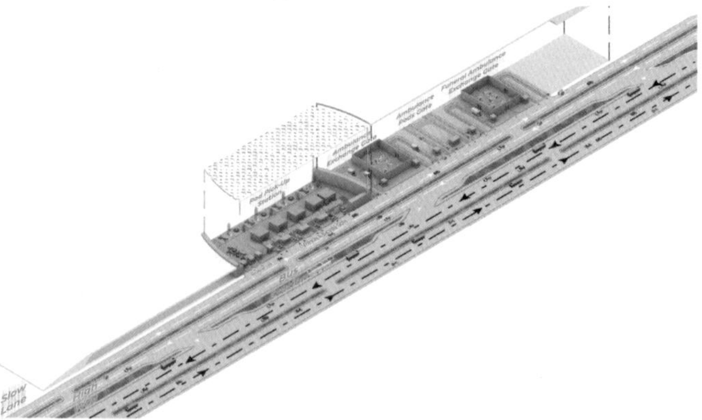

轻轨站 Monorail Station

地铁站 Metro Station

建筑设计 Architecture Projects

平面图 Layout

综合医院
General Hospital

中心医院
Central Hospital

模块
Legos

流行病医院
Pandemics Hospital

康复中心
Rehabilitation

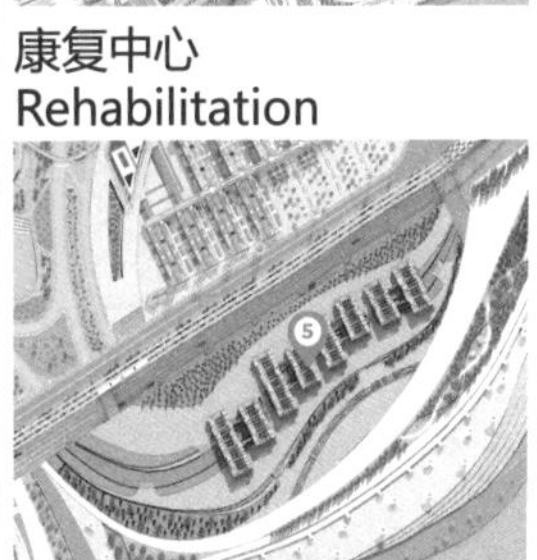

教育设施
Educational Building

1 & 4.综合医院 & 流行病医院 General Hospitals & Pandemic Hospitals

总平面 Masterplan

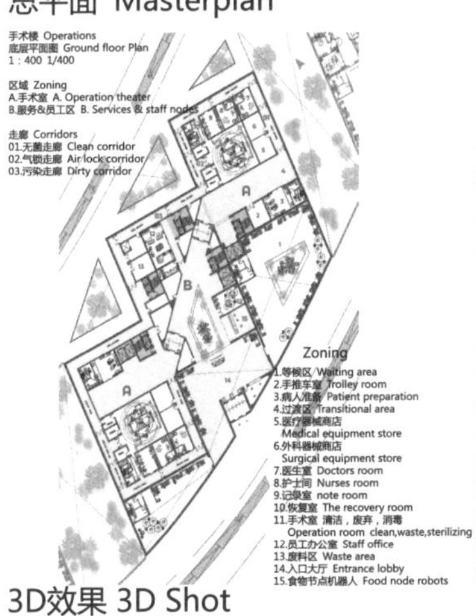

分区 Zoning

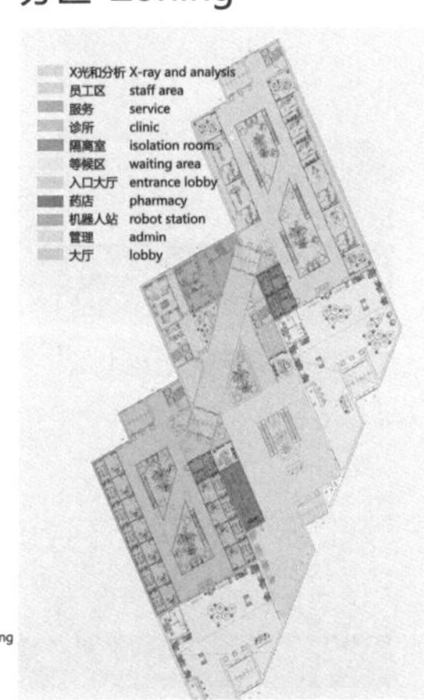

剖面 Section

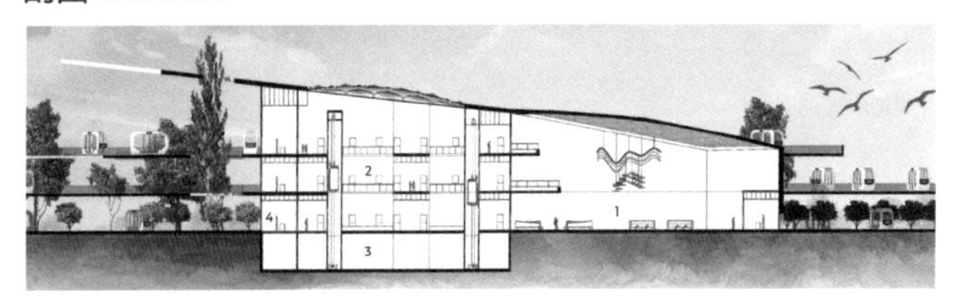

剖透视 Elevation

3D效果 3D Shot

3D效果 3D Shot

2. 中心医院 Central Hospital

总平面 Masterplan

专项设计 Zoning

剖面图 Section

立面图 Elevation

3D效果 3D Shot

3D效果 3D Shot

3. 乐高医院 Legos Hospitals

总平面 Masterplan

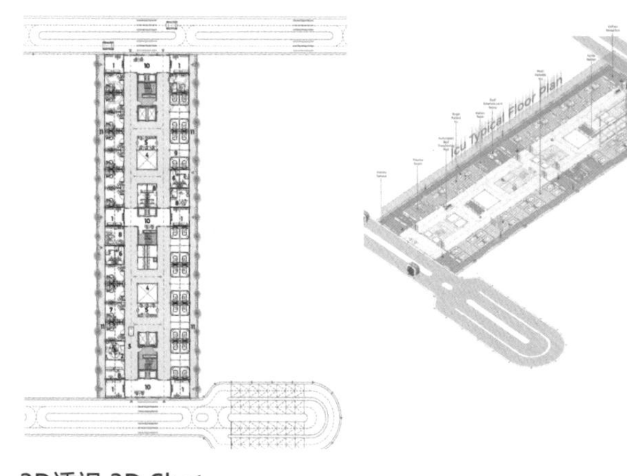

分区 Zoning

剖面图 Section

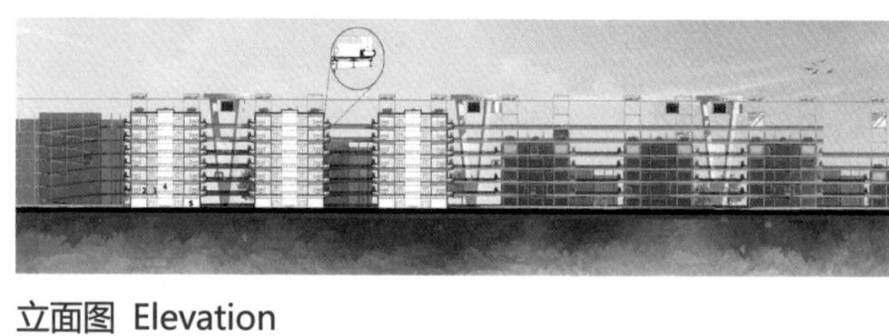

立面图 Elevation

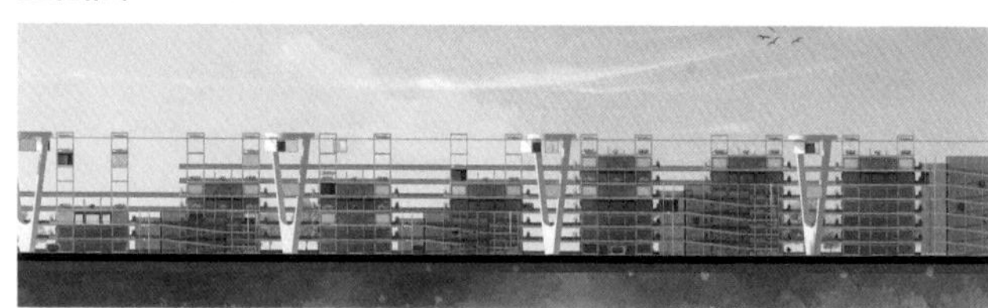

3D透视 3D Shot

3D透视 3D Shot

4. 康复中心 Rehabilitation Center

总平面 Masterplan

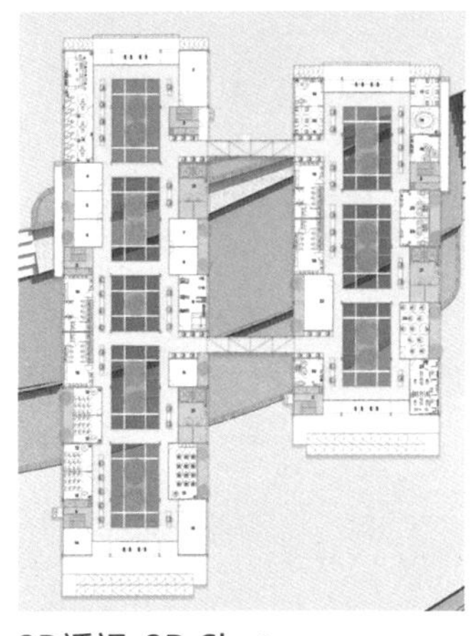

绿化 Greenery

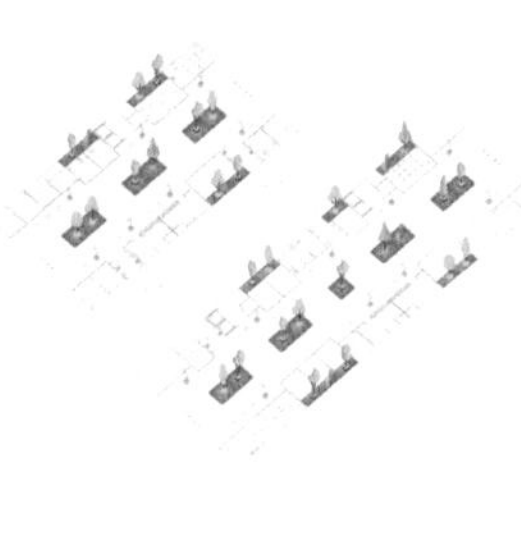

剖面图 Section

立面图 Elevation

3D透视 3D Shot

3D效果 3D Shot

5. 教育中心 Educational Center

总平面 Masterplan

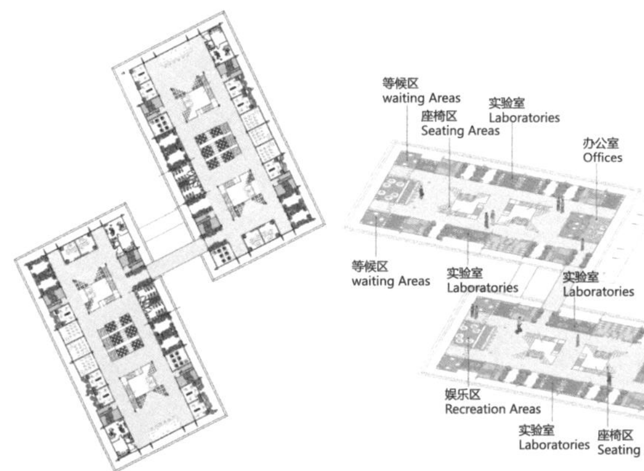

分区 Zoning

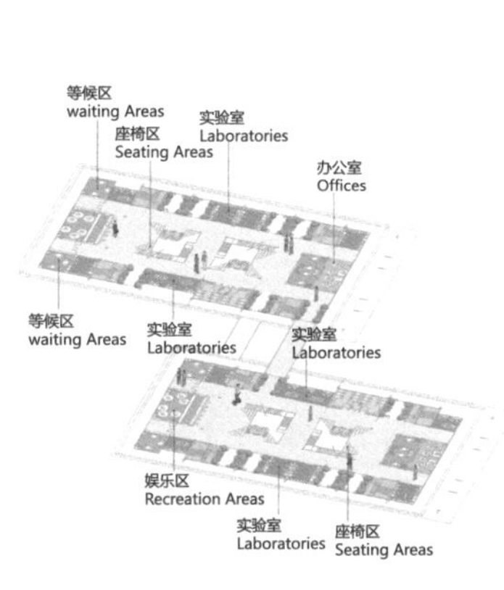

剖面图 Section

立面图 Elevation

3D效果 3D Shot

3D效果 3D Shot

邻里社区
Neighborhood Model

“如果我内心快乐，那么无论我住在何处，我都如同身在天堂。”

"If I am happy inside, then I live in paradise no matter where my residence is."

在考虑洪水、洪涝等现有限制的条件下，设计一个可步行、可骑行、能够抵御自然灾害和人为灾害的社区。

Designing a pandemics resilient neighbourhood, walk able, recyclable neighbourhood resilient to natural and man-made disasters taking in consideration existing constrains such as floods, waterlogging, etc.

设计分析 Design Analysis

分区 Zoning

社区具备自给自足所需的服务和工作人员。

The neighbourhood has services and daily human needed to be self-sufficiency.

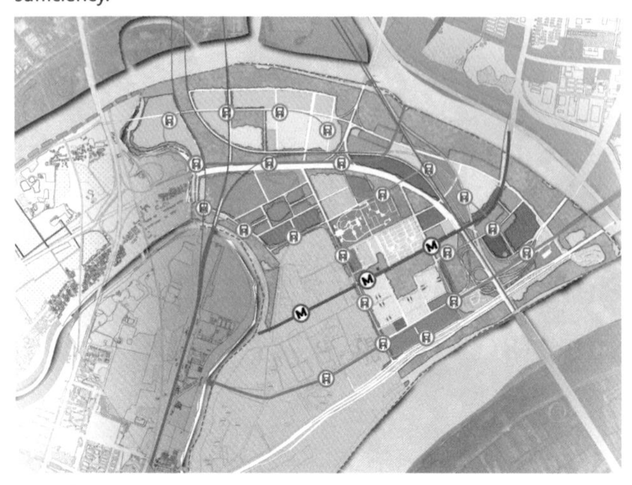

道路等级 Road Hierarchy

主要道路包括公共交通和连接主要道路和地方道路的次要道路。

Main roads contains public transportation and secondary roads which connect between main roads and local roads.

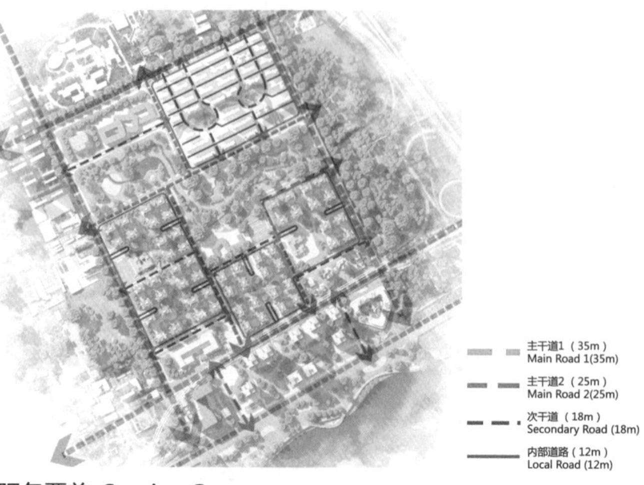

公共交通 Public Transport

用地铁串联整个区域，以避免外部环境同区域内部隔离开来。

Monorails are used to connect the whole districts in case of isolating from all the surroundings and internally.

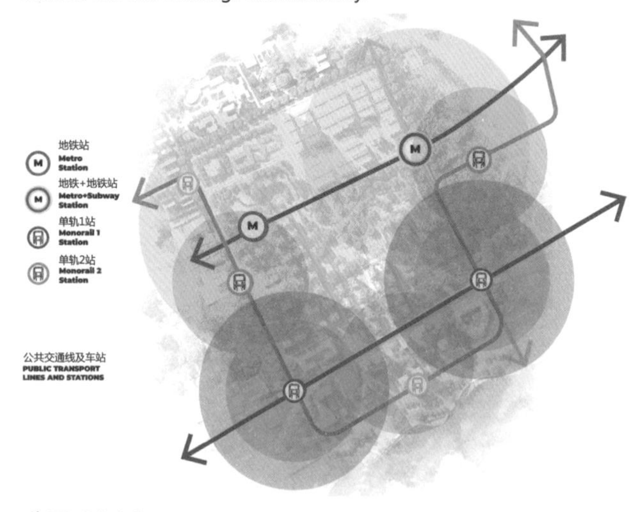

服务覆盖 Service Coverage

最大步行距离，高中700米，小学500米，街区设施150~300m。

Maximum walking distance: high school: 700m, elementary school: 500m, block facilities: 150~300m.

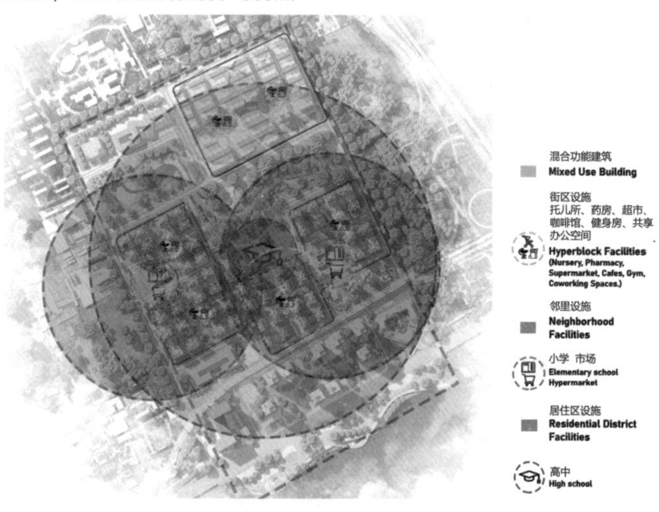

分区 Divisions

每个居住区由一个或两个街区组成，且每个街区功能各异。

Each neighbourhood is consists of one or two blocks. Each block is provided with its facilities.

景观分区 Landscape Zoning

为保护树木不被砍伐，所有不需要建筑的活动区域都置于本就有植被的地块。

All activities that don't require structure were held on areas with existing trees to prevent cutting the trees .

鸟瞰图 Aerial View

流行疾病研究 Pandemic Studies

Case 0 & A

情况概述：整个地块都是安全的，并没有被污染。
应对流行疾病的解决方案：整个地块都被保护起来了，没有恐慌的必要。

Description
The whole project is safe as there is no contamination in it.
Pandemic Solution
There is no need to panic as the whole project is protected from outer.

Case B-1

情况概述：地块外部被污染，但内部是安全的。
应对流行疾病的解决方案：通过在地块外围的主要道路上设置医疗隔离点将整个地块与外部隔离开来。

Description
The surrounding of neighbourhood is contaminated but inside is clear.
Pandemic Solution
Isolating the whole neighbourhood by making medical nodes on the main roads surrounding the neighbourhood.

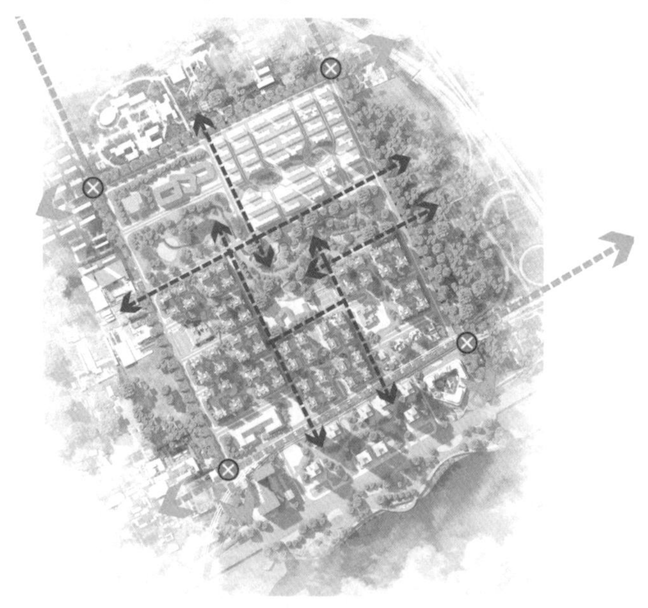

Case B-2

情况概述：地块内有一个感染区。
应对流行疾病的解决方案：将当地的公立学校设置为隔离病房，以此隔离地块内被污染的区域。

Description
The surrounding of neighbourhood is contaminated but inside is clear.
Pandemic Solution
The neighbourhood has a contaminated zone inside so we need to isolate it. Local public school turns into isolation unit.

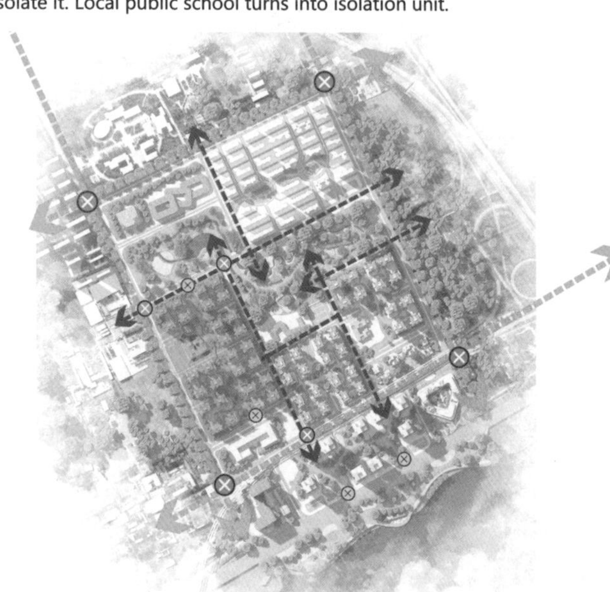

Case B-3 and Case C

情况概述：整个地块都被污染了。
应对流行疾病的解决方案：在所有主干道和人行道上设置医疗隔离点并隔离各区域，禁用绿廊步道，将学校用作隔离单元。

Description
The whole neighbourhood is contaminated.
Pandemic Solution
All zones must be closed to trap the infection by using medical nodes at all main roads & main pedestrian paths & forbidden the walkway in green corridor- public school turns into isolation unit.

景观 Landscape

平面图 Layout

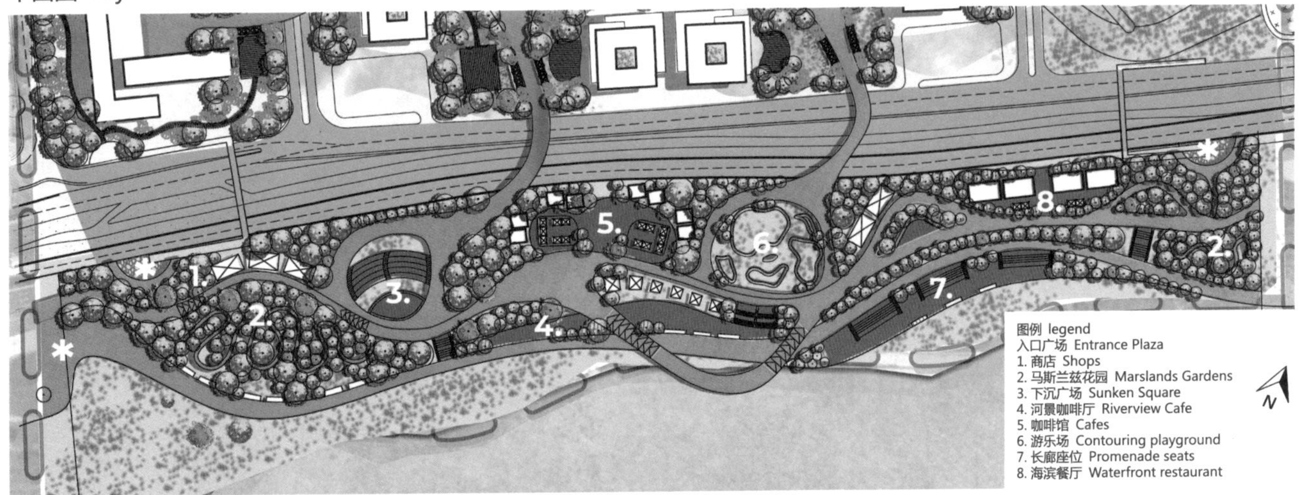

中心公园 Central Park

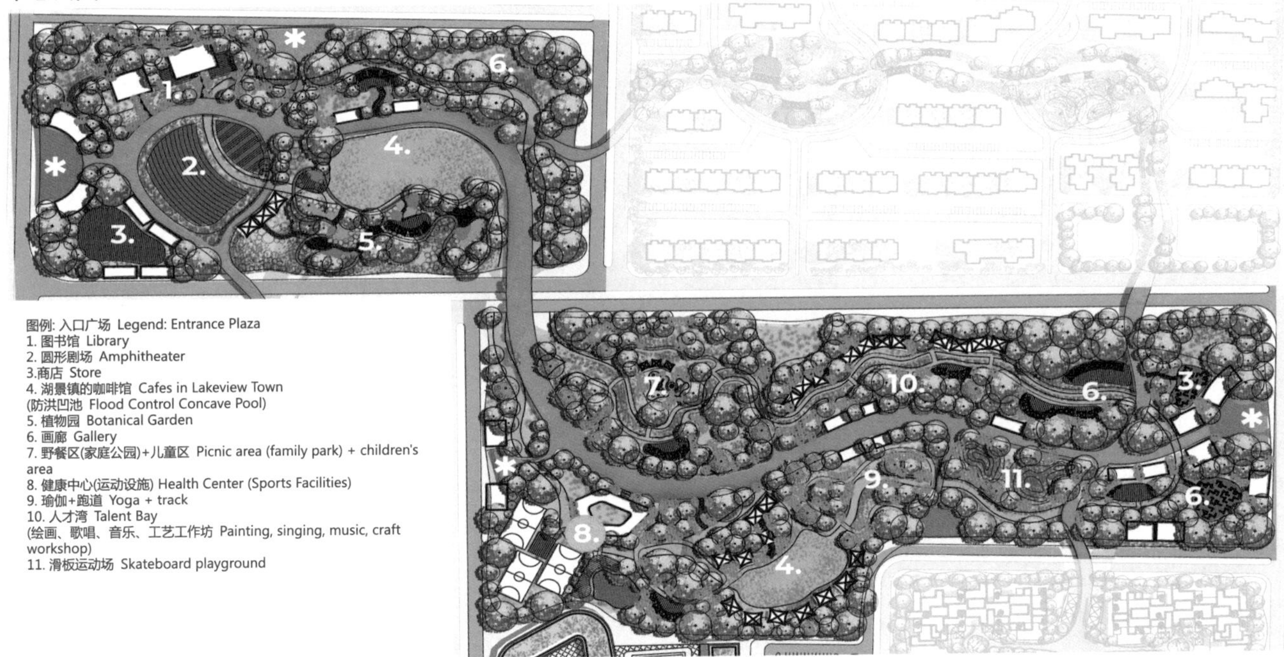

洪水墙 Flood Wall

用于应对洪水的可扩展地下防洪墙。

To help keep the project land resilient against flood, our design proposes an expandable underground flood wall in case of floods.

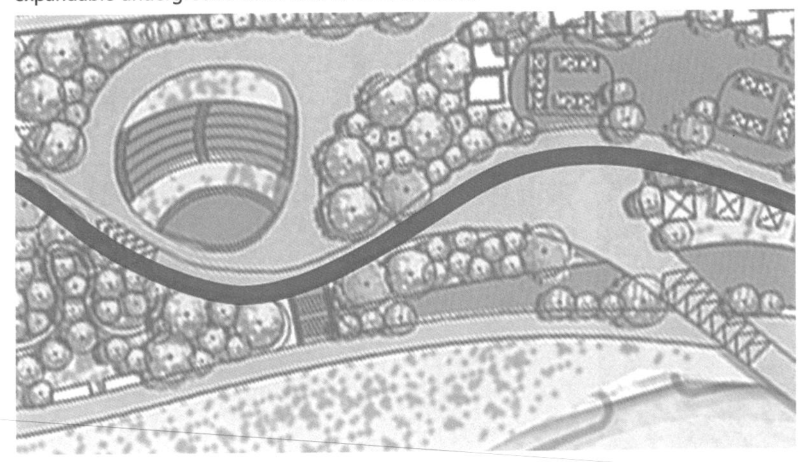

监测点 Check-up Nodes Pandemic

在区域间的出入口设置医疗监测站点。

Medical nodes are placed at each entrance/exit from one zone to another.

建筑 Architectural

总平面 Master Plan

1.复合型居住区
Modular Residence

2.学校
School

3.出租单元
Rented Units

4.健康单元
Wellness Center

鸟瞰图 Aerial View

5.酒店综合体
Hotel Complex

6.紧急设施设置
Emergency Settlement

7.中心
Headquarters

1.复合型居住区 Modular Residence

在武汉，单身人群和小型家庭的平均居住单元为50~60m^2。
住宅楼：构建面积为50m^{2}和100m^{2}的特定模块。
出租单元楼：构建提供给员工的面积为50m^{2}和75m^{2}的特定的模块(5米×10米)。
Since the average residential unit in Wuhan for single users and small families is 50~60 sqm.
Residential Building: Using specific module with area 50sqm and 100 sqm.
Rented Units Building: Using specific module (5m×10m) and area 50sqm and 75 sqm for employees.

总平面 Master Plan

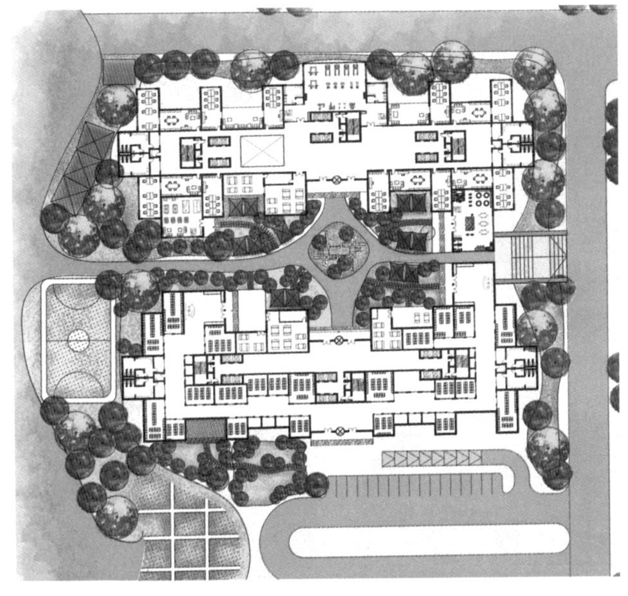

剖面图 Section

社区内沿人行道的半公共区域(包括儿童游乐场、售货亭、画廊，以及车间、城市农场、自行车停车场、自动取款机)。

Semi public zones inside the neighbourhood along pedestrian path (kids playground, kiosks, galleries and workshops, urban farms, bikes parking, ATM machines).

典型建筑平面 Typical Floor Plan

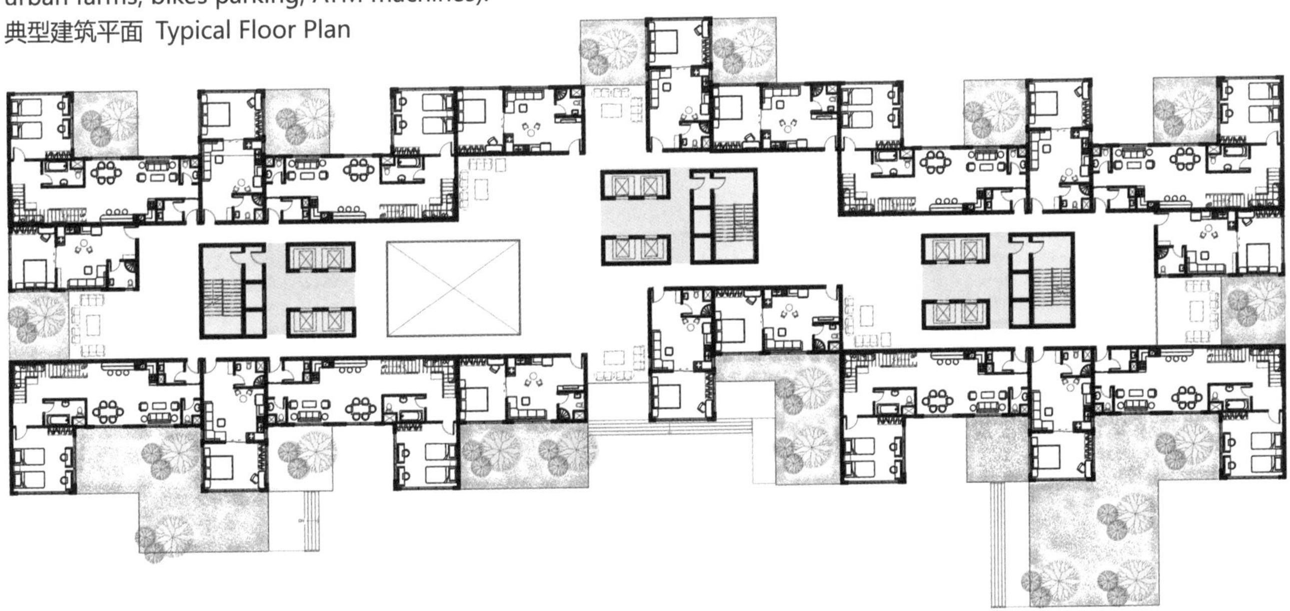

立面图 Elevation

2. 学校 School

学校平面由重复的单元组成，在流行疾病下可以转换成用于隔离的房间。其建筑的外观设计则采用了中国风主题。

The School has a repeated unit in the plan which can be converted into isolated room in pandemic case and using Chinese themes in the exterior design of the Building.

3D效果 3D Shot

立面图 Elevation

剖面 Section

平面图 Master Plan

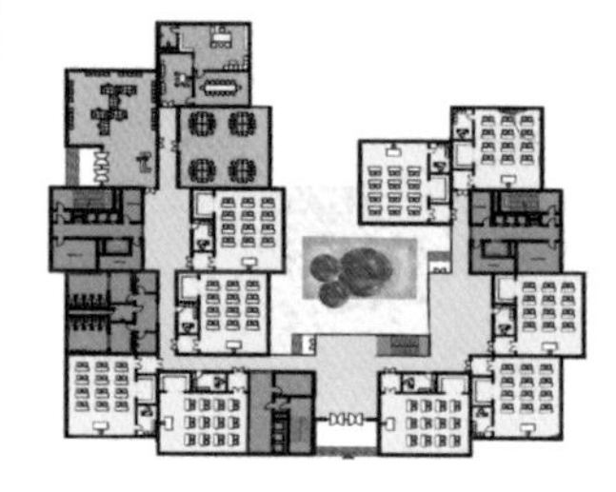

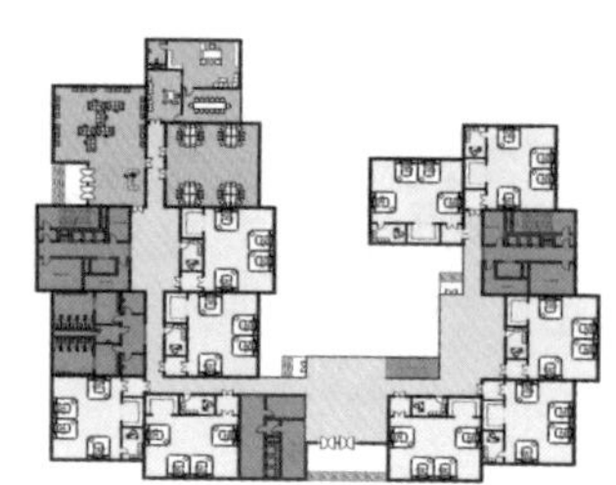

3.出租单元 Rented Units

方形建筑将通过增加模块化单元在竖向上进行空间扩展，通过搭建临时建筑在景观区域实现水平扩展。

The squared buildings will expand vertically by adding modular unit and horizontally on landscape by fast construction buildings.

平面图 Master Plan

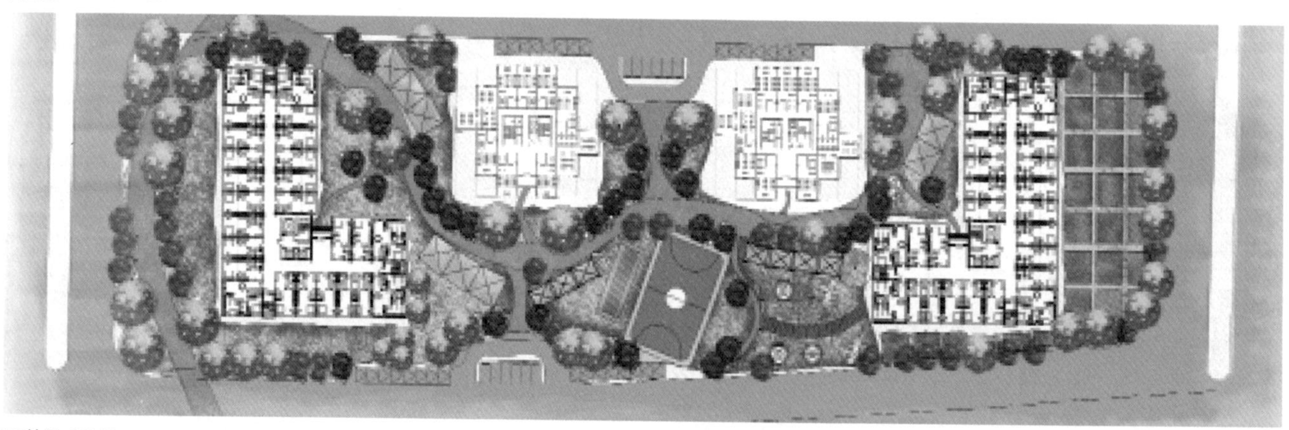

3D效果 3D Shots

立面图 Elevation

剖面图 Section

4. 健身中心 Wellness Center

从社交距离来看，中央公园内的健康中心可以满足周边居民的活动需求。建筑物的屋顶分为两层，低层用于浴室和教室等小空间，高层则用于儿童活动区这样的大空间。

The wellness center is located in the central park to complete the activity cycle in the neighborhood taking in consideration social distance .The roof of the building has two levels, a low level for small spaces such as bathrooms and classrooms and high level for large spaces such as kids area.

平面图 Master Plan

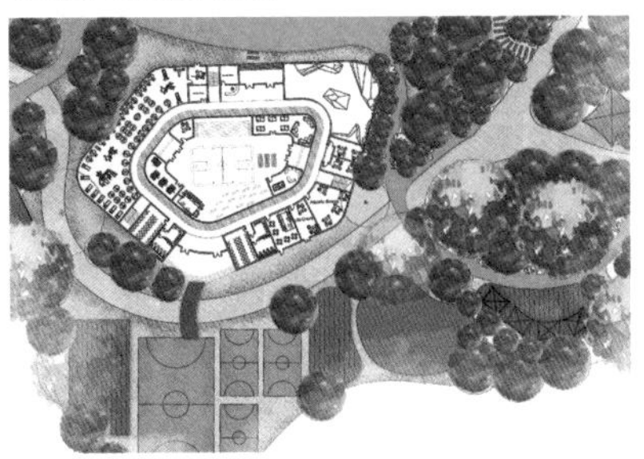

3D效果图 3D Shots

5. 总部 Head Quarter

该设计的灵感来自于八卦图，虚实相交，并通过传统与现代材料对比，构建了大量可供员工休憩的休闲空间和绿色露台。

The design is inspired by Yin and Yang symbol as presence of solid and void in mass creating staff leisure space and green terraces and contrast between tradition and modern materials.

平面图 Master Plan

3D效果图 3D Shots

立面图 Elevation

6. 紧急避难 Emergency Settlements

该行政管理单元可以包含8米×8米的模块化单元，可以垂直扩展，也可以通过在周围景观上建立一个6米×5米的单元化建筑来水平扩展，该单元可以传输到医疗中心。

The administrative can be extended vertically as it contains modular unit 8m×8m or horizontally by establishing a unitized building on the surrounding landscape with a unit 6m×5m which can be transmitted to medical hub.

平面图 Master Plan

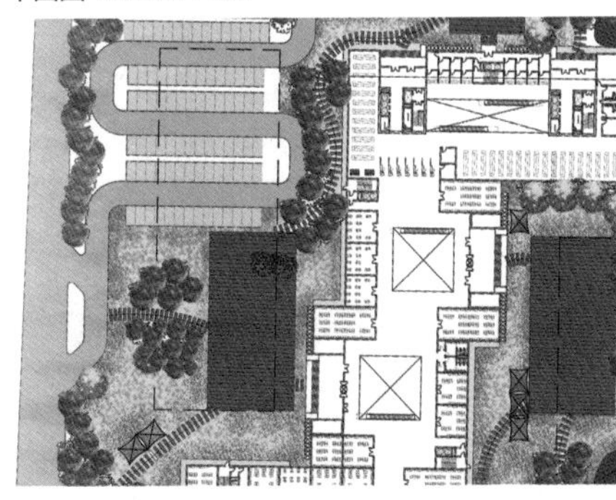

3D效果图 3D Shots

剖立面图 Elevation and Section

7. 酒店 Hotel

该建筑采用阶梯式设计，利用其空间作为绿色区域，投下绿荫，并通过镂空使其自然通风，以此提供良好的用户体验和可持续性。

The building is stepped to make use of the spaces as green areas ,3 buildings with subtraction to allow natural ventilation through the buildings and a great green shade across the 3 buildings for a great user experience and sustainability.

总体规划图 Master Plan

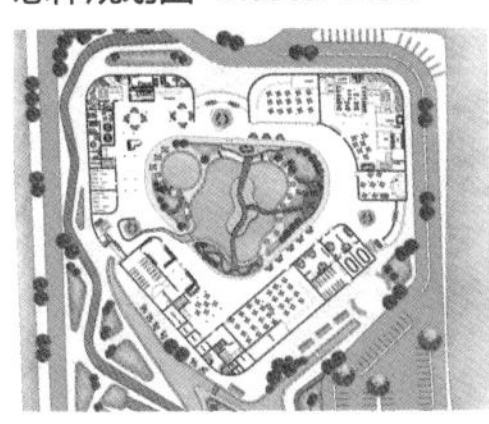

效果图 3D Shot

剖面 Section

效果图 3D Shot

奥林匹克城市
Olympics City

这座城市将作为一个磁场，为全世界各民族奥运城市的主要地区充当磁铁的两极，在这两极之间将是来自周围不同国家观众的磁场流。

The city will act as a magnetic field for all the nationalities around the world where the main zones of the Olympic city acts as the poles of the magnet and between this poles will be the flow of magnetic field that will be spectators from different countries around the world.

设计目的 Design Goals

该项目背后的主要分区理念是用两个主要区域（靠近地铁站的体育场和海滨的竞技场）带动其他区域发展。之所以如此布局，其主要目的是创建一个顺畅流动和连接的环流，以方便大批人群舒适地使用。

参数化工具最适合我们创建代表充电点之间最自然和理想路径的磁力线，这些正负电荷的力量基于空间层次结构给体育场和竞技场这两个复杂区域带来了最大的分布影响。

The main zoning idea behind this project is having two main major zones (stadium located near the metro station and arenas on waterfront) that attracts other zones towards them. The main intention with the concept of the layout is to create a circulation that flows and connects seamlessly in order to facilitate the large masses with ease and comfort.

The parametric tool was most suitable for us creating magnetic field lines representing the most natural and desired paths between charged points, the power of these positive and negative charges are based on space hierarchy which gives the stadium and arena complex zones the biggest effect on the distribution.

功能区划 Macro Zoning

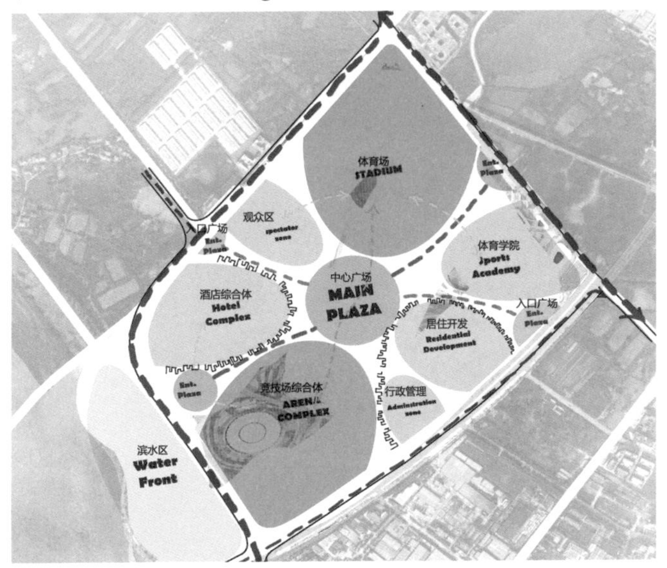

绿色空间 Green Spaces

剖透视 3D Section for Environmental Concerns

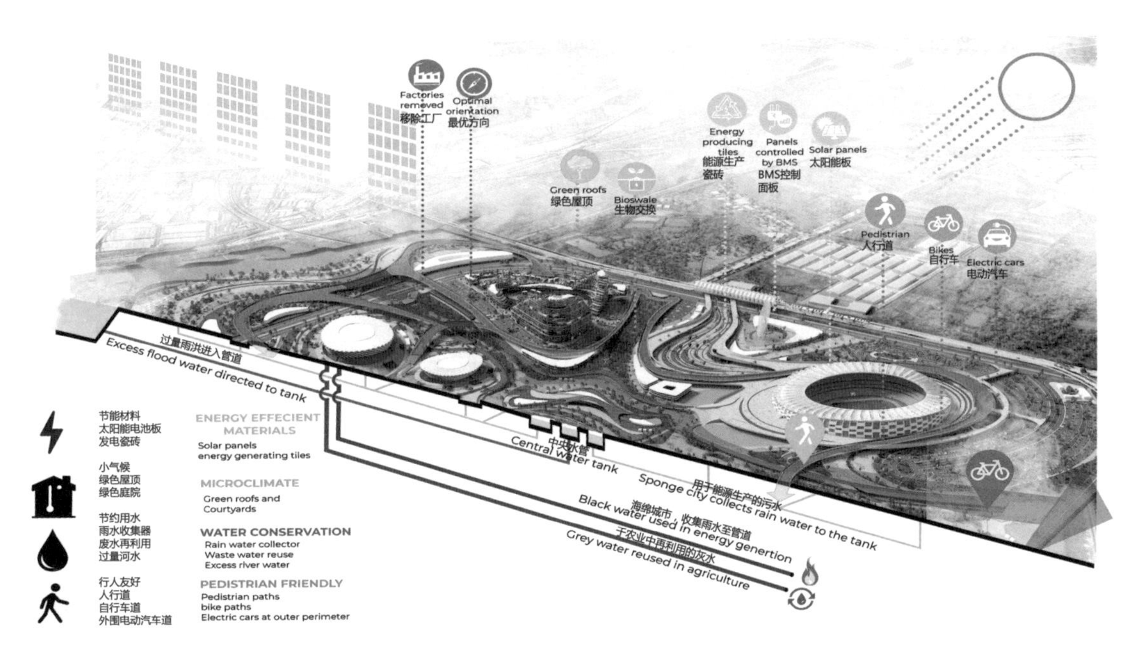

道路层次 Roads Hierarchy

移动性Mobility

鸟瞰图 Aerial View

流行疾病下的弹性情景 Pandemic Resilience Cases

该项目已经过消毒，运行良好，并采取了预防措施以防止流行疾病在游客之间传播。在广场及建筑物入口使用与移动应用程序相连的排队系统，以调整特定的访客人数。地铁站、停车场的检查站、项目入口使用消毒门。同时设计有动态调平独立座位的运动场和体育场。在主广场使用标识减少和分散人群。运动区运行良好，受水节点和道路节点的控制，尤其是从区域边缘进入，并在案例C中完全封闭。

The project is disinfected and functioning well and taking precautions to prevent the spread of pandemic between visitors. Using Queuing system linked to mobile application to adjust specific number of visitors at the plaza and buildings entrances. Checkpoints at the metro station and parking, use disinfection gates and sanitizer gates and sanitizer gates at project entrances. Dynamic levelling Isolated seating are designed at sports arena and stadium. Signs are used to decrease crowds in main plaza. The sports zone functions well with controlled access by water nodes ,and road nodes specially from the district edge, and completely closed at Case C.

如果流行疾病在区外，而区域内未受感染，则在道路、水源、山麓入口和地铁站设立检查站和节点，检查是否有人携带疾病，如果有，则由自主医疗车送往最近的陆地紧急安置点。

Where pandemics is outside the district and our district is uninfected, checkpoints and nodes are established at road, Water, piedmont entrances and metro station to check whether people are carrying the disease or not and if so, they are sent to the nearest emergency settlement to the land by autonomous medical cars.

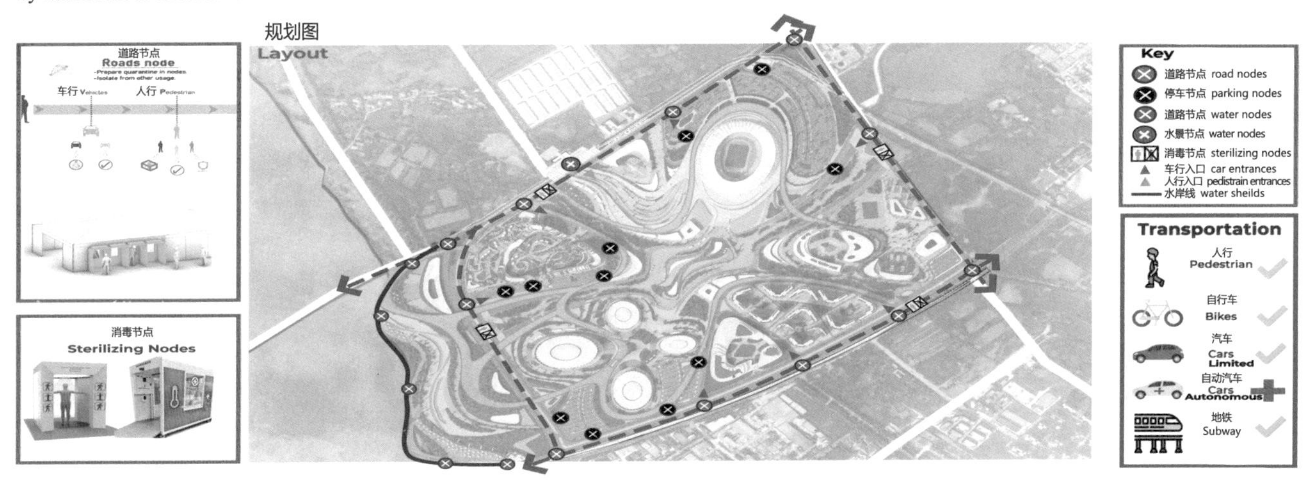

在流行疾病最严重的情况下，土地受到感染，所有入口严格关闭，设有检测节点，地铁站关闭，除自动医疗车和救护车外，任何车辆不得进入。主广场被改造成野战医院，土地建筑单元通过使用智能家具被改造成医疗单元。自动无人机用于对开放空间进行消毒，并向医疗单位提供医疗设施。

Where pandemics reached its worst case and the land is infected, all entrances are strictly closed with detection nodes and metro station is closed, no cars are allowed to enter except autonomous medical cars and ambulances. Main plaza is converted to a field hospital and land buildings units are converted to medical units by using smart furniture.Automated drones are used to sterilize open spaces and to deliver medical utilities to medical units.

风景园林设计 Landscape Design

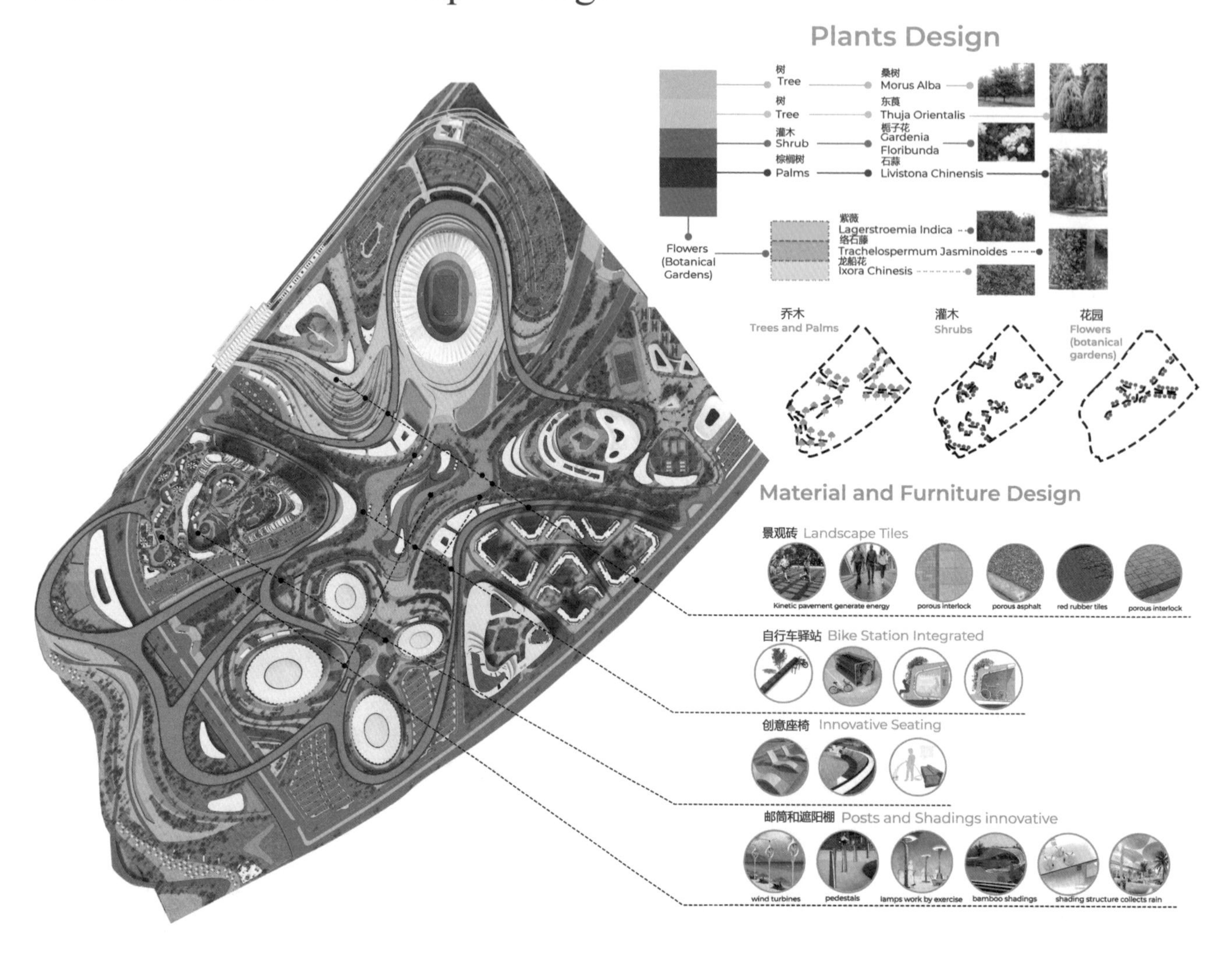

场地剖面 Site Section

效果图 3D Shots

滨水设计 Waterfront Design

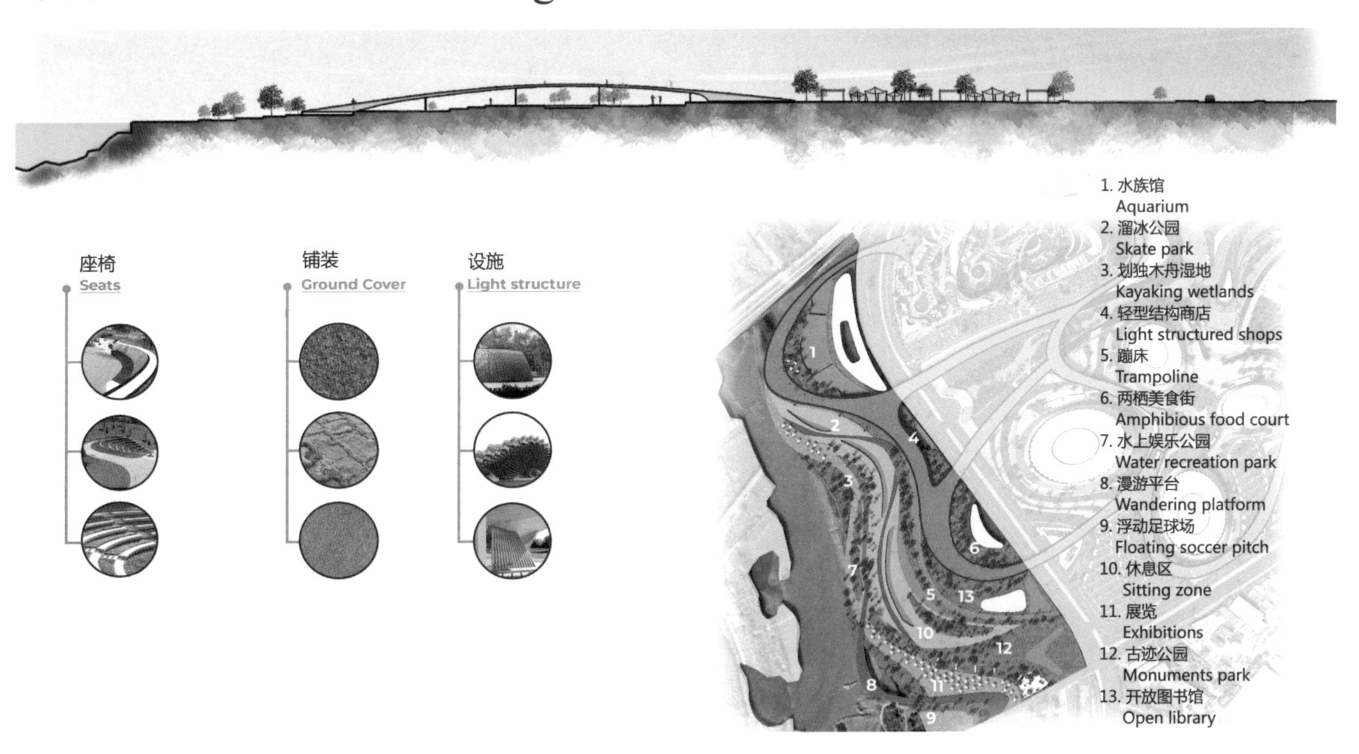

建筑项目 Architectural Projects

规划图 Layout

鸟瞰图 Aerial View

1.体育馆 Stadium

2.综合竞技场 Arena Complex

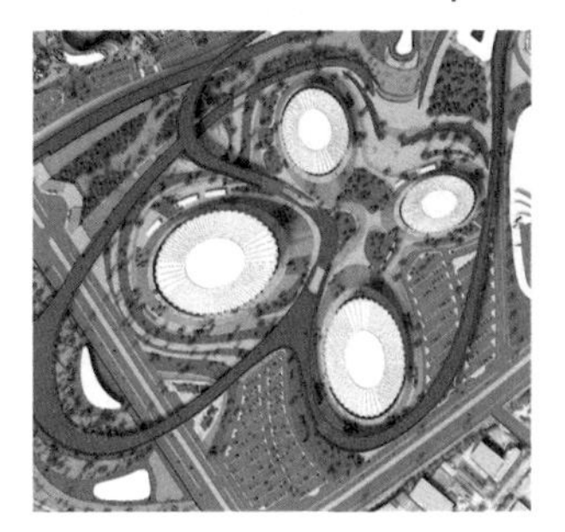

3.体育学院 Sports Aacademy

4.多功能酒店 Hotel complex

1. 体育馆 Stadium

一层平面 Ground Floor Plan

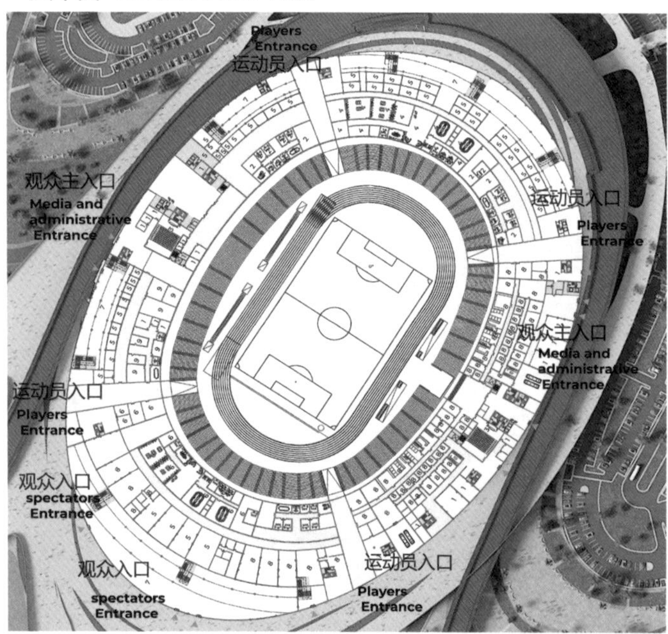
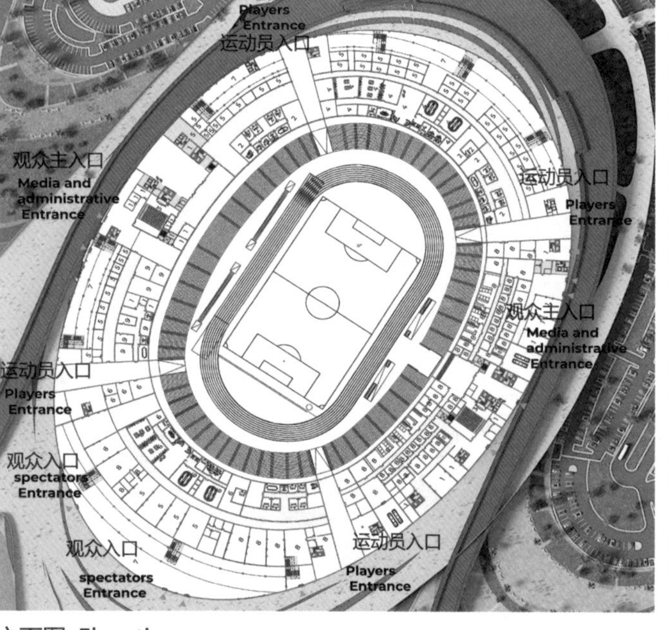

场地效果图 3D Shot

内部效果图 Interior Shot

立面图 Elevation

剖面图 Section

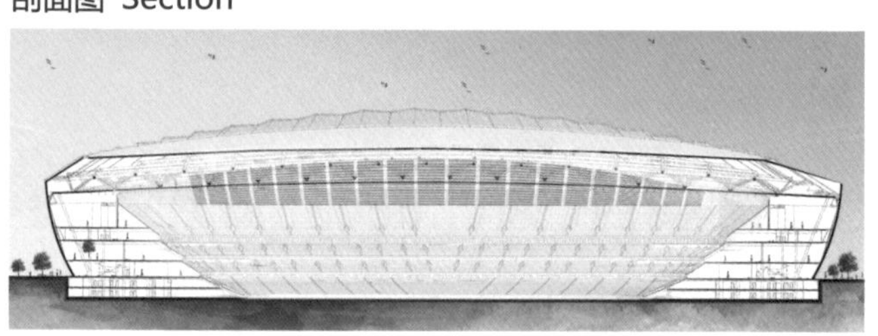

2. 综合竞技场 Arena Complex

一层平面 Ground Floor Plan

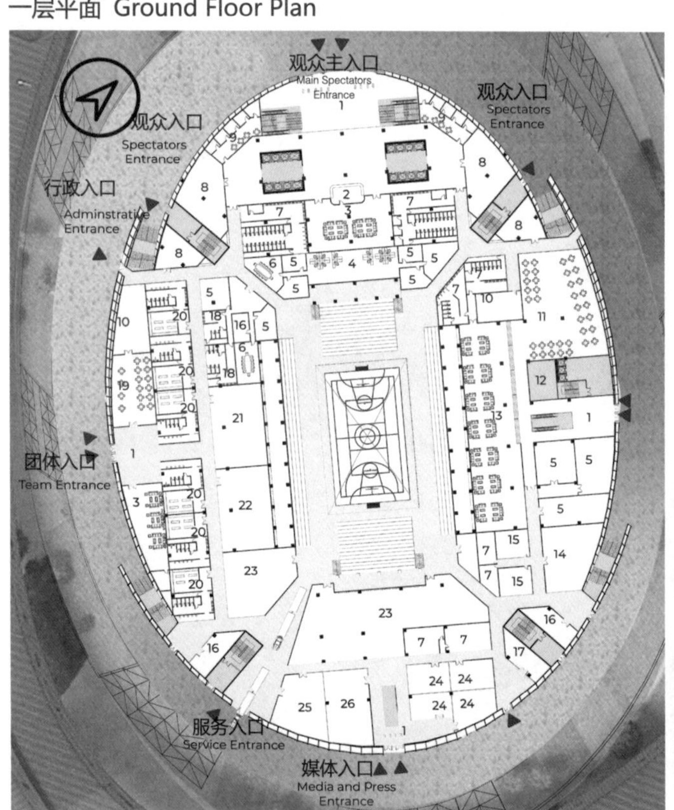

剖面图 Section

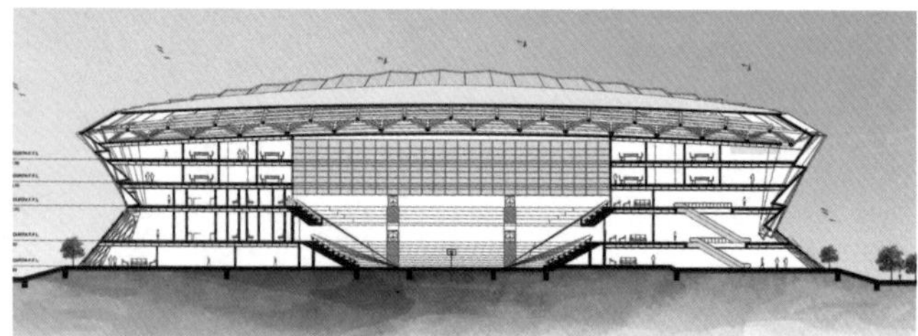

效果图 3D Shot

3. 体育学院 Sports Academy

平面图 Ground Floor Plan

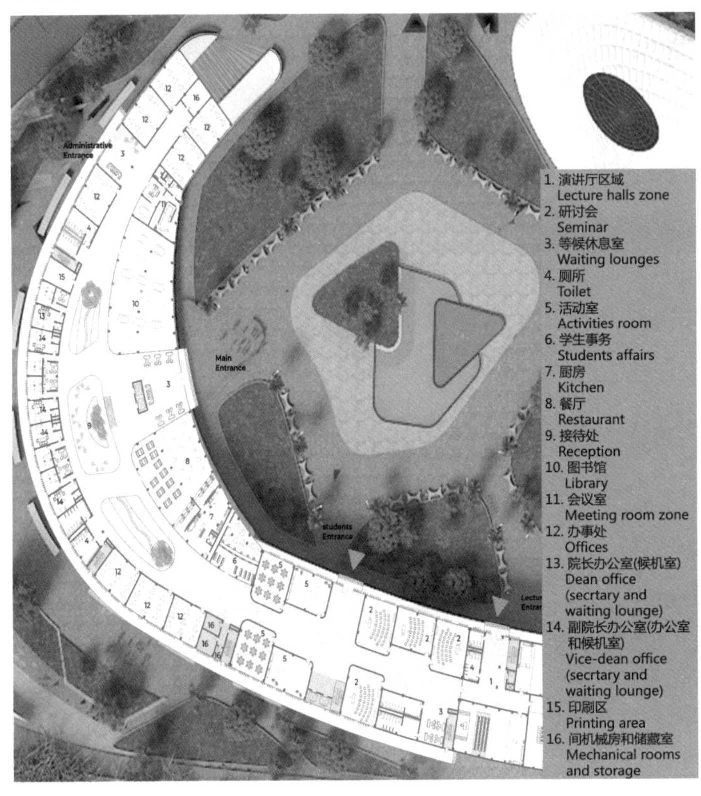

立面图 Elevation

剖面图 Section

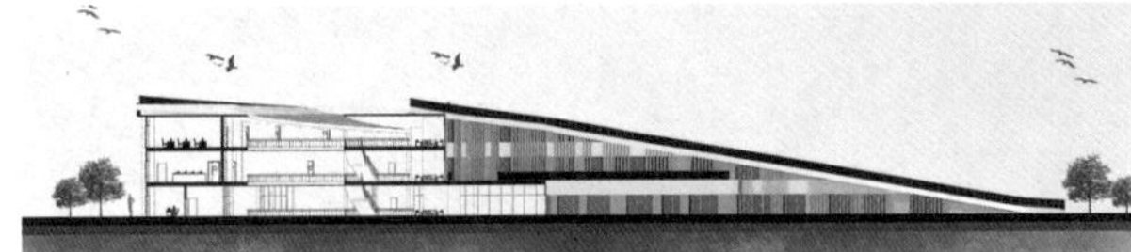

效果图 3D Shot

4. 多功能酒店 Hotel Complex

一层平面图 Ground Floor Plan

效果图 3D Shot

剖面图 Section

内部效果图 Interior Shot

鸟瞰图 Aerial View

生态工业园

Eco-Industrial Park

新的工业革命是尊重环境的工业。

The new industrial revolution is about industry with respect to the environment.

“在生态工业园区内，具有一个集成式的生产循环系统，其中一家工厂的产品是另一家工厂的原材料。”

通过以下开发方式，尽可能在环境保护和牟取经济利益之间达到平衡：

增加废物处理方法和措施；

用绿地形成缓冲区域；

使该工业园区成为区域内经济可持续发展的重要支柱。

“Eco-industrial park with an integrated production loop where one plant’s production is the raw material for another plant.”

The balance between environmental protection and economic benefits shall be achieved as far as possible through the following development methods:

Increase waste treatment methods and measures;

Use green space to form buffer area;

Make the industrial park an important pillar of regional economic sustainable development.

专项设计 Design of Special Topic

分区 Zoning

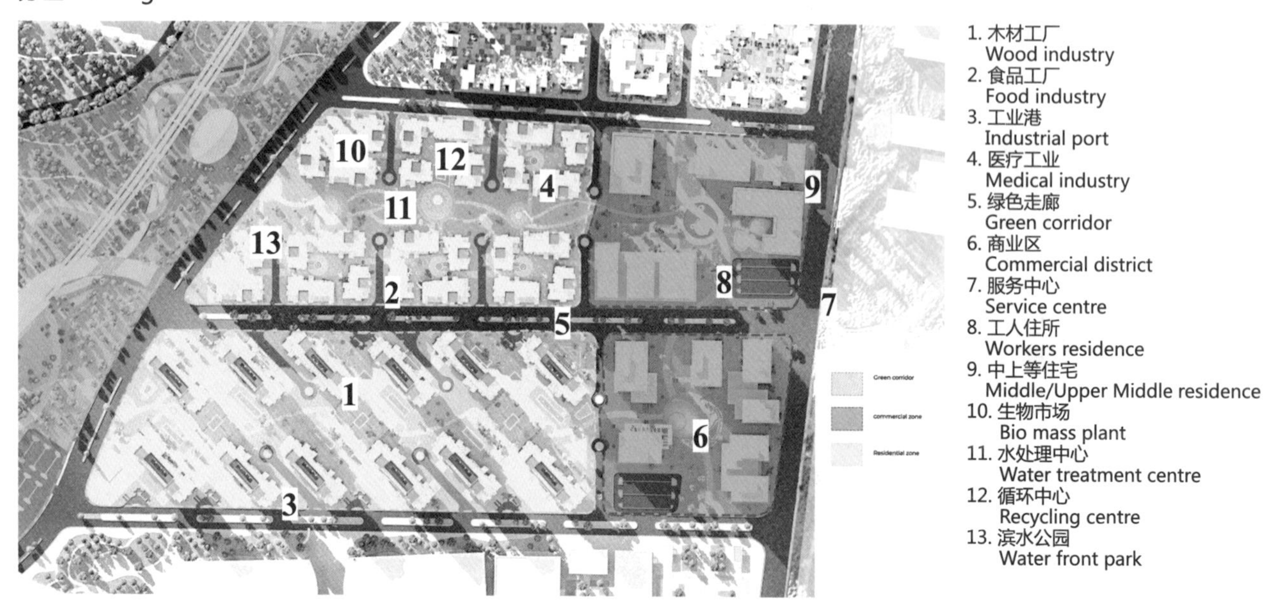

道路层次 Roads Hierarchy

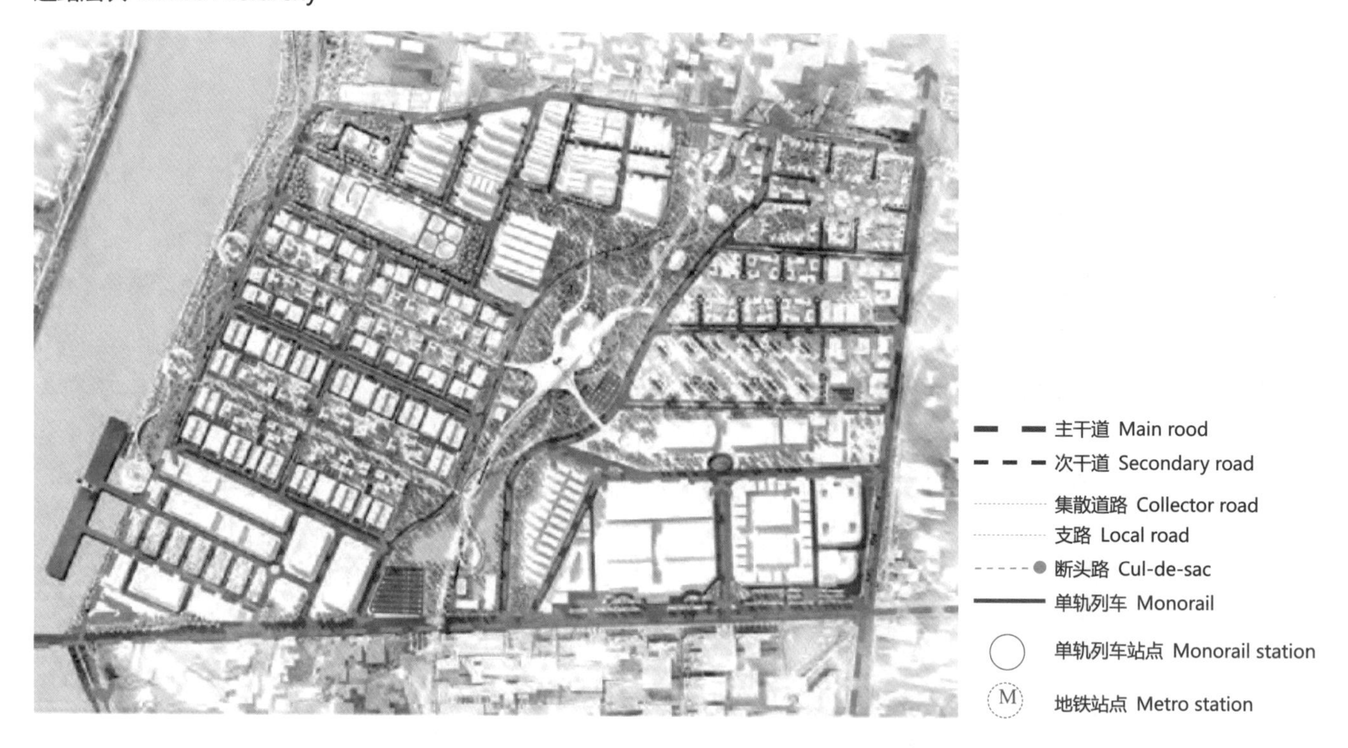

布局分为两个部分。滨水部分由食品、木材簇和工业港口组成。住宅和商业区部分靠近外部道路，中间部分是绿色走廊，绿色走廊将所有部分连接在一起并提供行人网络，形成绿色网络，其中包括与各个区域相关的许多活动，例如下沉广场、家庭公园、喷泉广场、城市农场、森林深处等。

The layout is separated into two parts. The water front part consists of food, wood cluster and industrial port. The residential and commercial part is near to external road, the intermediate part is the green corridor which creates a green network connecting all the parts together and providing a pedestrian network, it includes many activities related to each zone such as sunken plaza, family park, fountain plaza, urban farms, deep forest, etc.

服务设施 Services

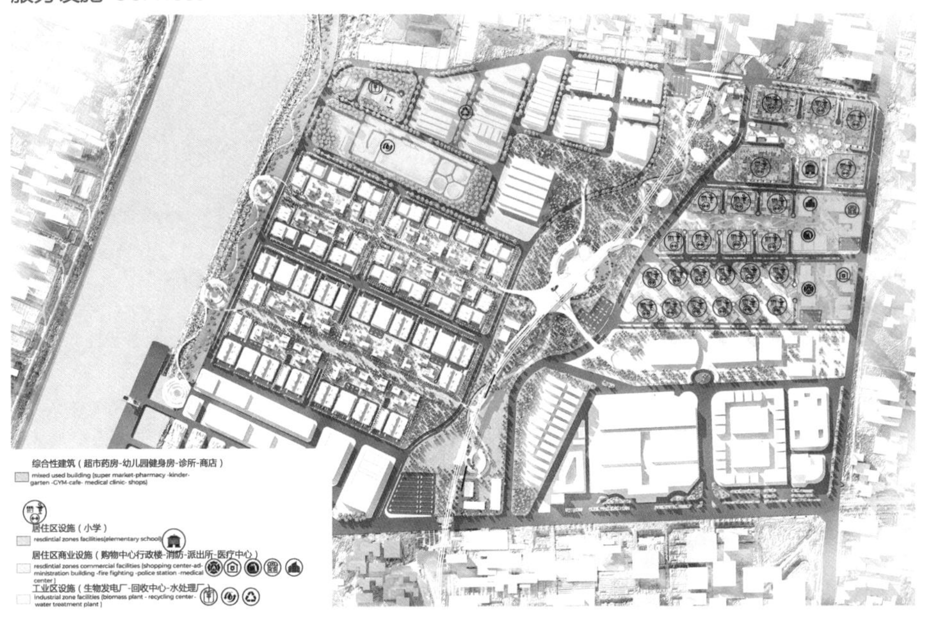

植物园 Botanical Gardens

空间层次 Space Hierarchy

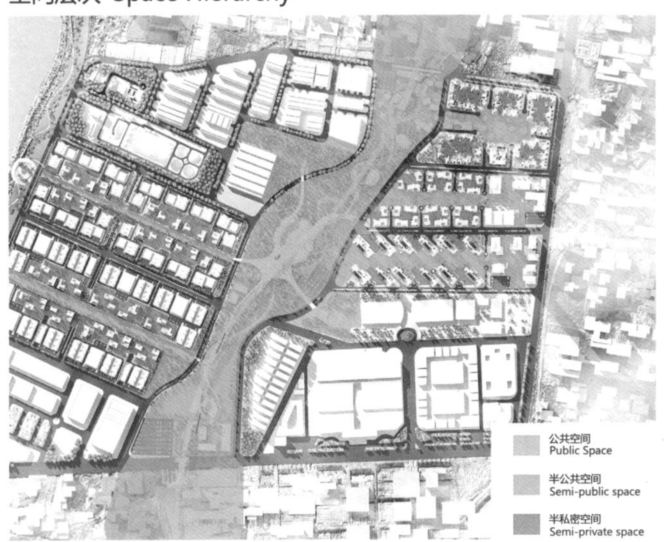

绿色廊道 Green Corridor

鸟瞰图 Aerial View

抵抗流行疾病的案例 Pandemic Resilience Cases

使用道路节点和水屏障将场地与其他区域分开。

所有的人必须留在自己的地方不离开，公共场所安排有特定的广场 。

感染者将被转移到医疗中心，密切接触者将被紧急安置。

The zone is separated from the other zones using road nodes and water shield.

All people will be required to stay in their places and not to leave it ,the people in public places are required to go to specific plazas.

The infected person will be transferred to medical hub and the contacted one will be held in emergency settlement.

在确定项目级别之前提出策略
Propose strategy before identifying the level of cases

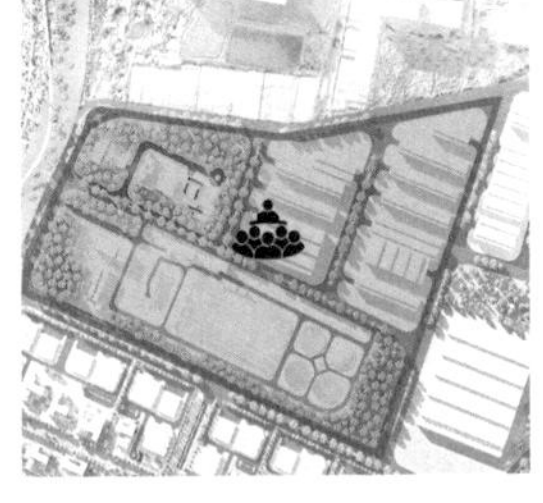

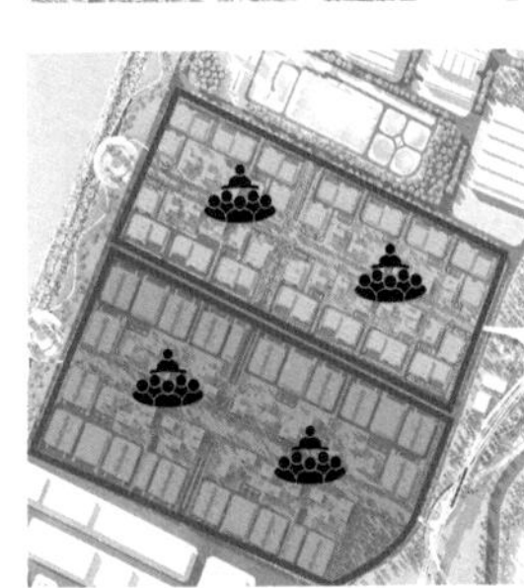

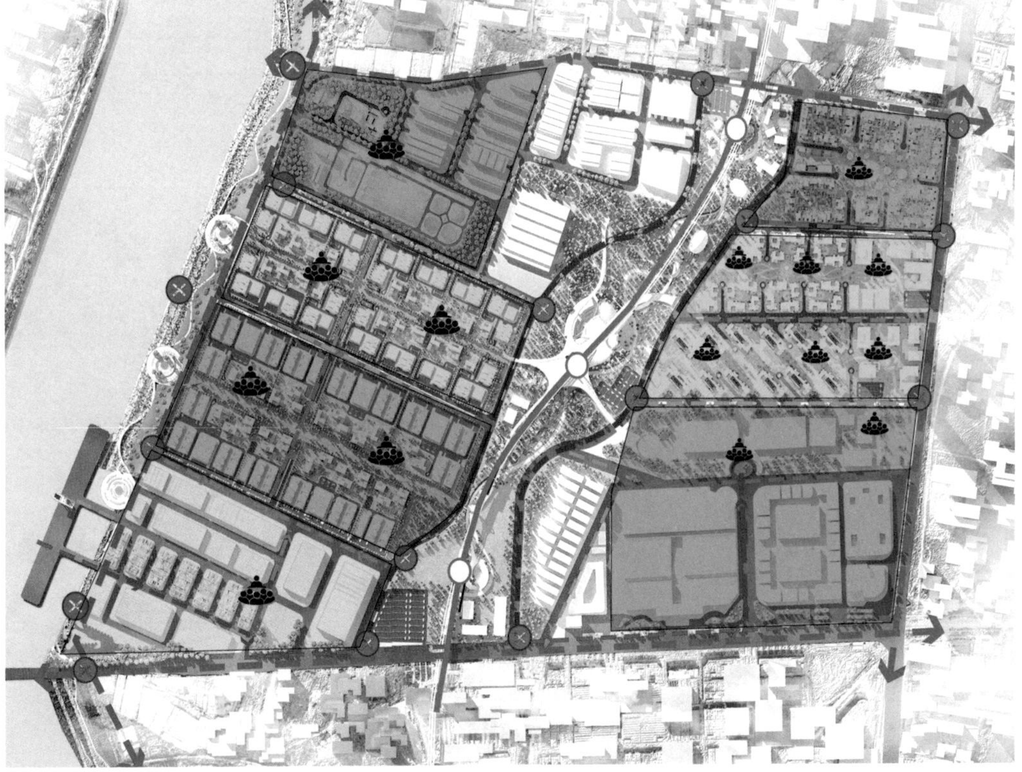

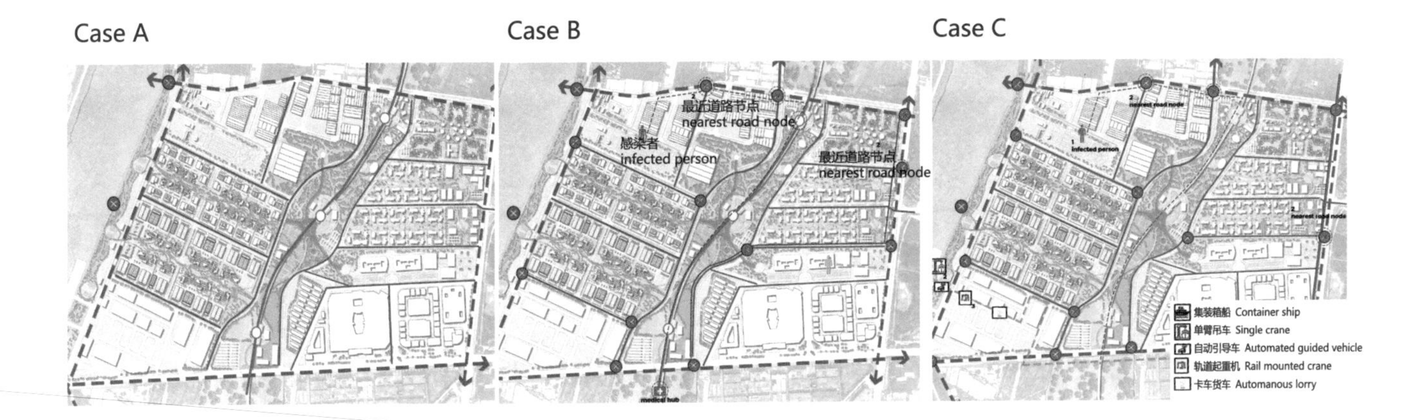

风景园林设计 Landscape Design

绿色廊道活动 Green Corridor Activities

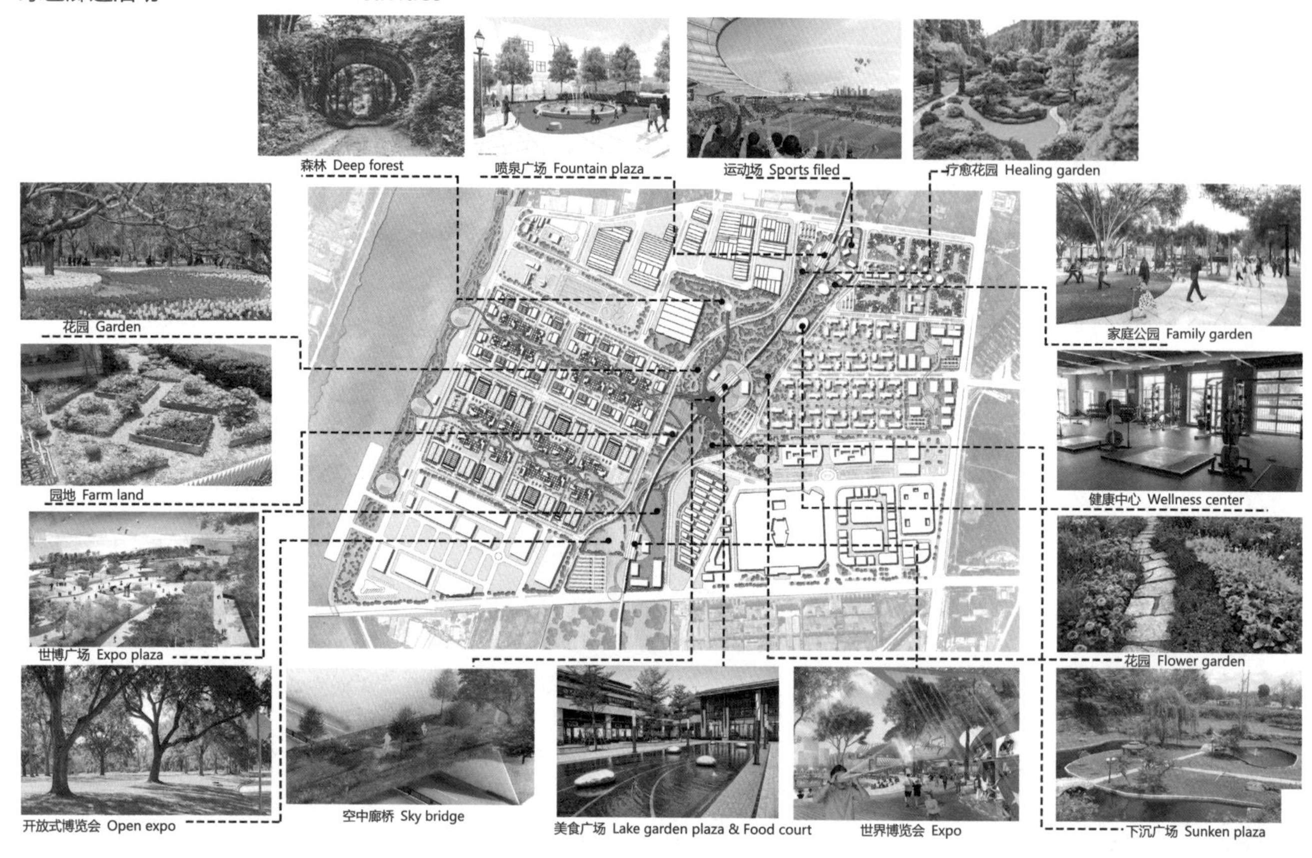

绿色廊道区域 Green Corridor Zoning

1. 停车 Parking
2. 服务 Services
3. 运动场 Sports field
4. 天桥 Sky bridge
5. 家庭公园 Family park
6. 健康中心 Wellness center
7. 森林 Deep forest
8. 花园 Flowers garden
9. 植物园 Botanical garden
10. 下沉公园 Sunken garden
11. 农场 Farm lands
12. 世博会 Expo
13. 花园草坪 Garden lawn
14. 疗养花园 Healing garden

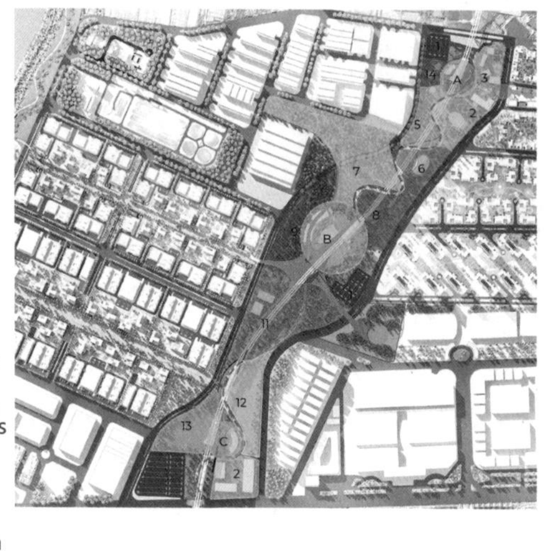

主广场设计 Main Plaza Design

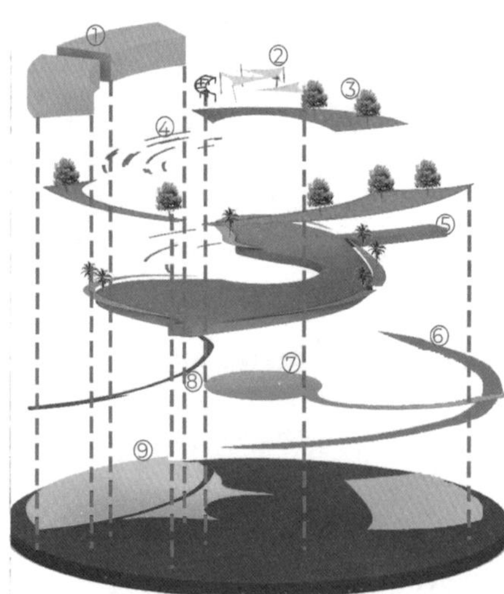

① 餐厅大楼 Food court building
② 遮阳构筑 Shading elements
③ 绿色植物 Greenery
④ 绿色空间 Green space
⑤ 水景 Water feature
⑥ 绿色走廊 Green corridor
⑦ 浮动路径 Floating path
⑧ 美食广场 Foodcourt
⑨ 次干道 Secondry path

剖面图 Section

滨水剖面 Section at water front plaza

建筑设计 Architecture Design

平面图 Layout

1. 食品厂 Food Industry

2. 木材工业 Wood Industry

3. 中上等收入住宅 Middle/Upper-Middle Income Residence

4. 工人住宅 Workers Residence

1&2. 食品厂和木材工业 Food Industry and Wood Industries

布局 Layout

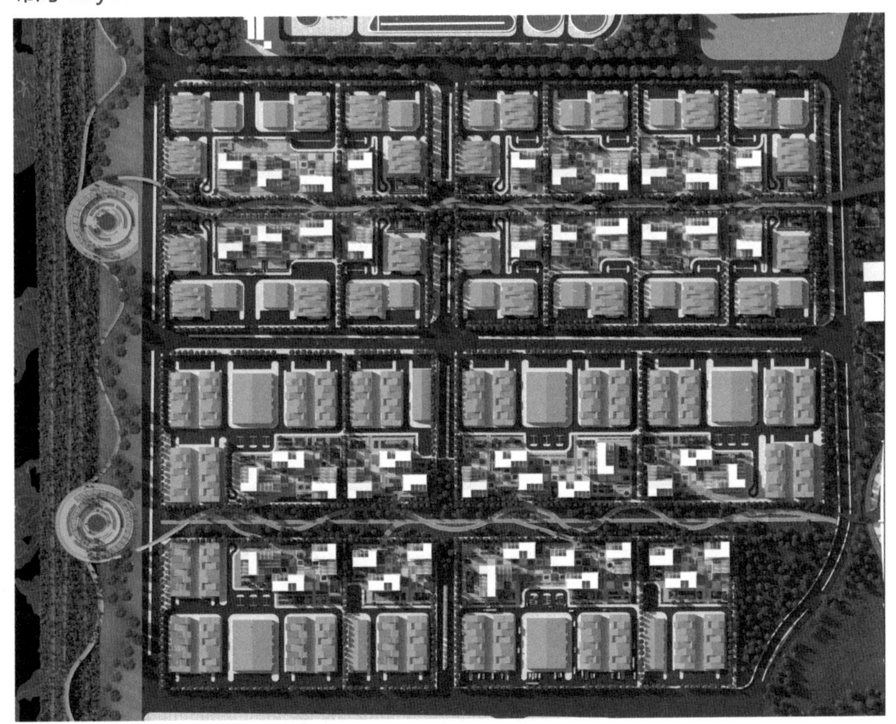

分区图 Zoning

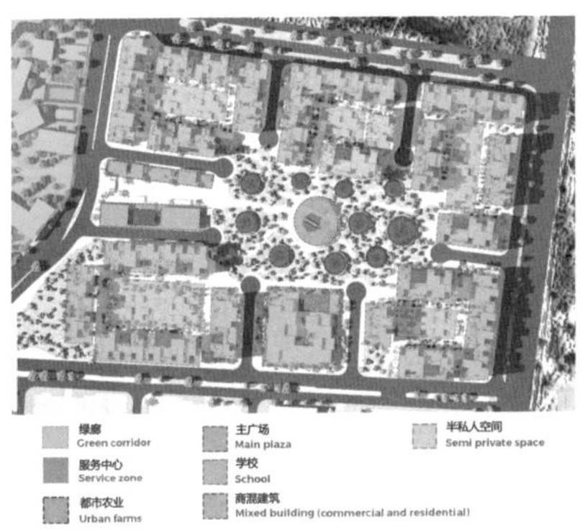

这个产业集群由许多基本单元构成。每个基本单元包含行政和研发建筑（位于绿色走廊分支上）以及工厂（位于外部道路）。

The cluster consist of several prototypes. Each prototype contains its administration & RD building (located on the green corridor branch) and the factories (on the external roads).

研发大楼 Research and Development Buildings

楼层平面图 Floor Plans

A型 Type A

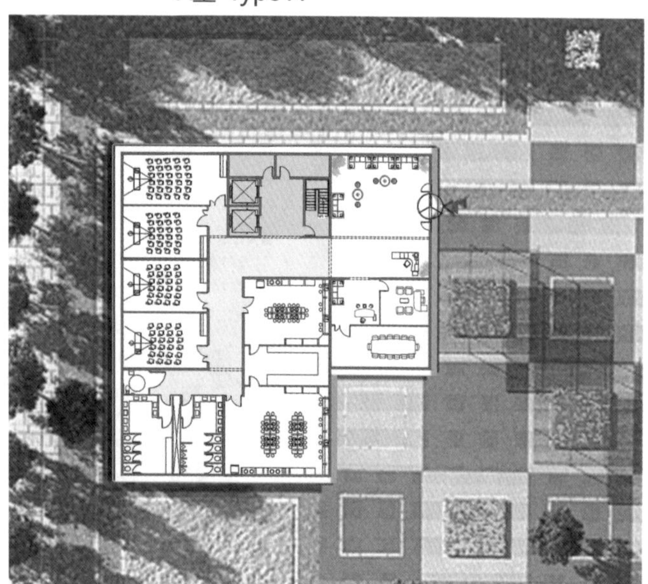

剖面图 Section

A型 Type A

立面图 Elevation

A型 Type A

产业集群 Industrial Cluster

楼层平面图 Floor Plans

A型 Type A

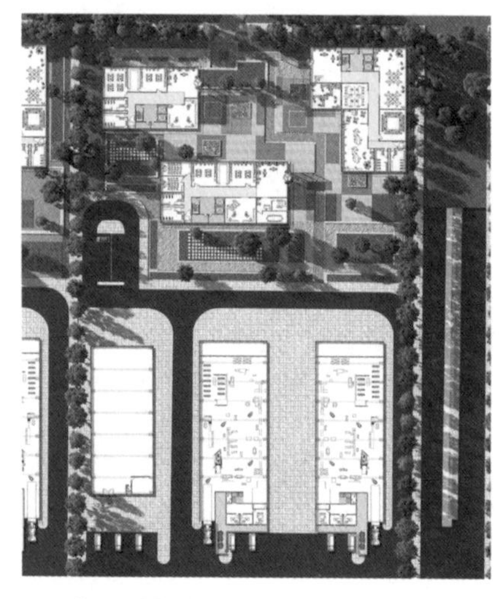

立面图 Elevation

A型 Type A

专项设计 Section

专项设计 Type A

效果图 Shots

3. 中上等收入住宅 Middle/Upper-Middle Income Residence

总体规划 Master Plan

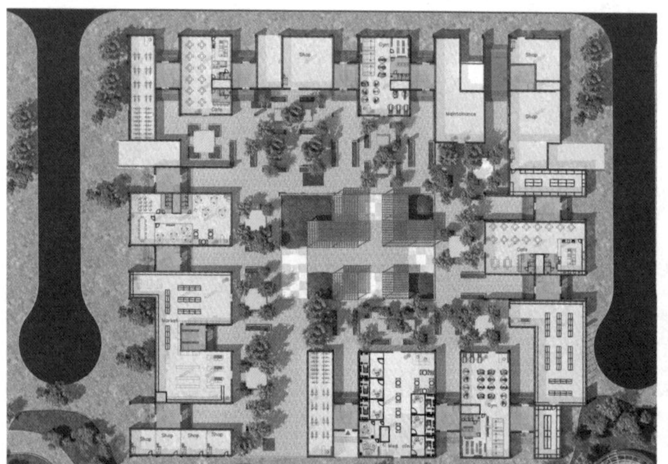

剖面图 Section

效果图 Shot

鸟瞰图 Aerial View

4.工人住宅 Workers Residence

总体规划 Master Plan

立面图 Elevation

效果图 3D Shot

剖面图 Section

文化廊道

Cultural Corridor

中国文化是中国的，也是世界的。

The Chinese culture belongs not only to the Chinese people but also to the whole world.

文化廊道内的建筑和城市空间能够带来一系列游憩体验，游客感受到强烈的舒适感或者不适感。这种体验感的实现需要经历 3 个阶段。在体验本项目时，来访者将按顺序或单独体验各个阶段。最终，对人类和地球上其他每一种生命形式的责任的种子将植根于每一位来访者的脑海。

The cultural zone through it's various architecture and urban spaces will provide a series of experiences designed to make the visitor at the end feel the grave consequences of even the simplest acts either positively or negatively. This experience is achieved through 3 phases and the user will pass through each of them either sequentially or separately throughout the whole project. Ultimately at the end the seed of responsibility towards humanity and every other living form on the planet in general will be planted in the minds of every visitor.

设计研究 Design Studies

新冠肺炎大流行告诉世界，我们最应该关心的是大自然，并让我们知道，我们只是大自然的客人。

长江的主要文化区向世界传授从这场流行病和全球危机中吸取的教训——人类的行为在影响自然前会影响人类自身，无论是不当的社会互动还是像洗手这样简单的事情。

最终，文化区的主要目的是让来访者意识到我们作为人类触发的无数定时炸弹，如果现在不拆除，我们所有人和子孙后代都将遭受严重痛苦。

The covid-19 pandemic taught the world that our main concern should be mother nature and showed us that we are just it's guests.

Ultimately the cultural zone's main aim is to make visitors aware of other countless time bombs that we as humans triggered and if not dismantled now all of us and future generations will suffer gravely.

主路径 Main Path

行人桥 Pedestrian Bridge

单轨路径 Monorail Path

道路层次结构 Roads Hierarchy

鸟瞰图 Aerial View

水体 Water Bodies

绿地 Green Bodies

鸟瞰图 Aerial View

弹性应对流行疾病的情景 Pandemic Resilience Cases

Case 0

预防总比治疗好。Case 0展示了在正常情况下，文化区扮演了最重要的角色，它通过互动的三个阶段告诉人们正常情况应该是怎样的，以提高人们对毁灭性问题的认识。

Prevention is better than cure. The cultural zone plays the most important role in Case 0 where everything is normal, it teaches people how the normal should be through an interactive three phases experience rising awareness on devastating problems.

Case A

当本项目范围内是安全的，但在其他地方出现零散病例的情况下，文化区的社交媒体团队开展针对疾病暴发地区的报道，提高本项目范围内民众对病情的认识，并报道世卫组织和其他可信机构的授权信息。

When the whole project is safe but there is a health problem rising in other places, use the cultural zone's social media team working on projects targeting the infected zones raising awareness of the current problem and use authorized information from WHO and other trusted institutes.

Case B1

当存在受感染区，但文化区不被视为隔离区时，部分隔离文化区，禁止来自疫区的人进入文化区，并且把在山中库存的防疫物资运送至病情较重的区域，通过道路节点控制文化区的入口。

When there is an infected district but cultural zone is not considered a quarantine zone, total lock down of the cultural zone for visitors and send aids from the storage area in the hill to infected zones and control the entrances of the cultural zone through road nodes.

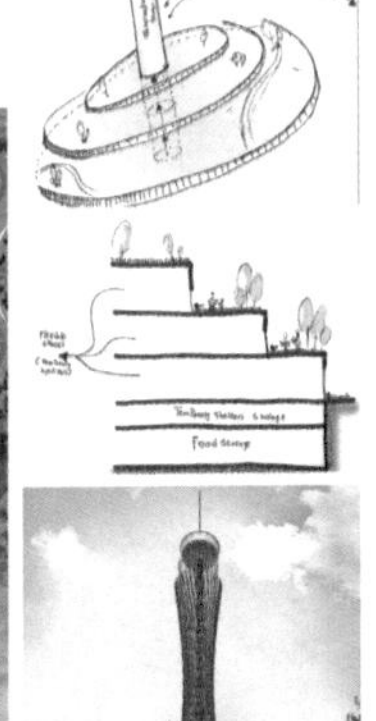

Case B2

当有受感染的地区时，文化区中的城市空间需要对区域进行全面封锁，为区域创造模块化结构隔离单位，使用山丘进行隔离，使用存储区进行隔离自给自足和作为无人机发射空间的观察塔。

When there is an infected district the urban spaces in the cultural zone are needed so total lockdown for the zone, creating modular structure for quarantine units, using the hill for quarantine, using the storage areas for self sufficiency and the observation tower as a launching space for drones.

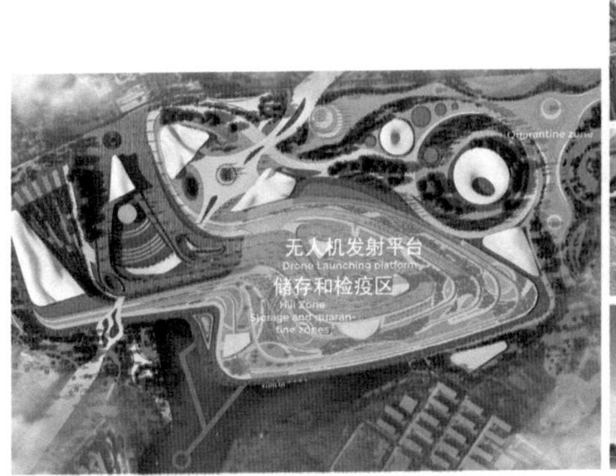

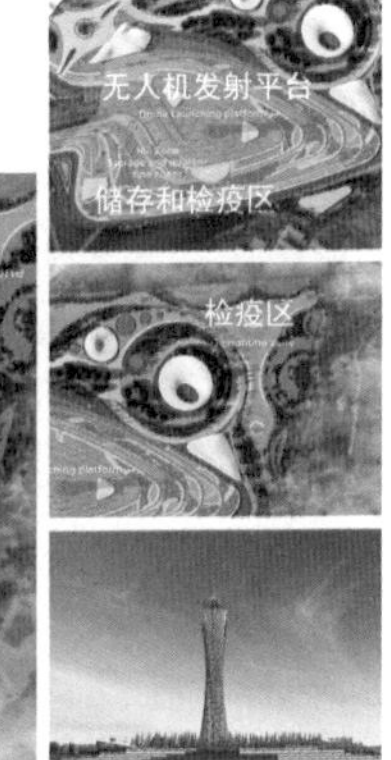

剖立面图 Sectional Elevation

风景园林设计 Landscape Design

详细效果 Detailed Shots

景观分区 Landscape Zoning

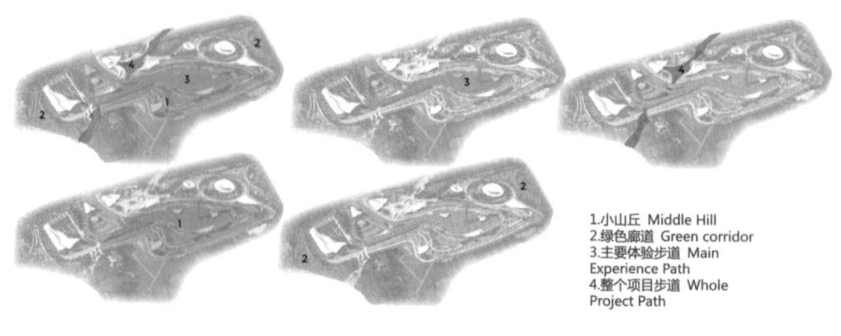

景观元素 Landscape Elements

景观小品 Landscape accessories

塔上拍摄的3D视图 3D Shot from Tower

从山地水平拍摄的3D视图 3D Shot from Hill

建筑设计 Architecture Design

总体规划 Masterplan

1.图书馆 Library

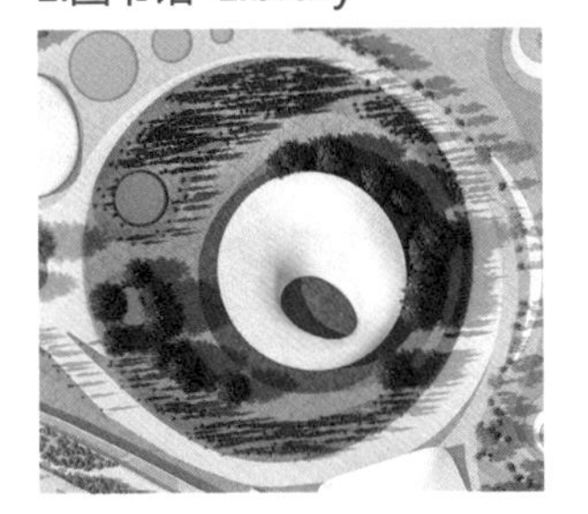

2.博物馆 Museum

3.文化中心 Cultural Center

4.观景塔 Observation Tower

1.图书馆 Library

总体规划 Masterplan

立面图 Elevation

3D效果图 3D Shot

3D效果图 3D Shot

2.博物馆 Museum

总体规划 Masterplan

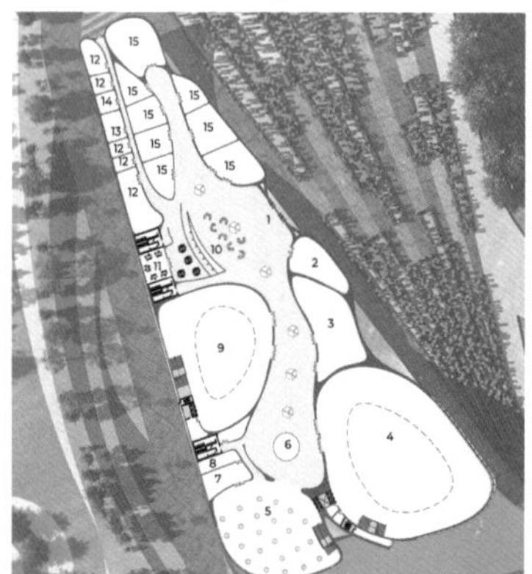

1. 入口大厅 Entrance lobby
2. 售票安保 Tickets/Security
3. 多功能厅 Multipurpose hall
4. 主展厅 Main exhibition hall
5. 餐厅 Dining hall
6. 开放商店 Open store
7. 厨房 Kitchen
8. 储物 Storage
9. 展览厅 Exhibition hall
10. 问讯台 Info desk
11. 管理处 Admin
12. 管理处 Admin
13. 恢复 Restoration
14. 储物 Storage
15. 展览 Exhibitions

3D剖面图 3D Section

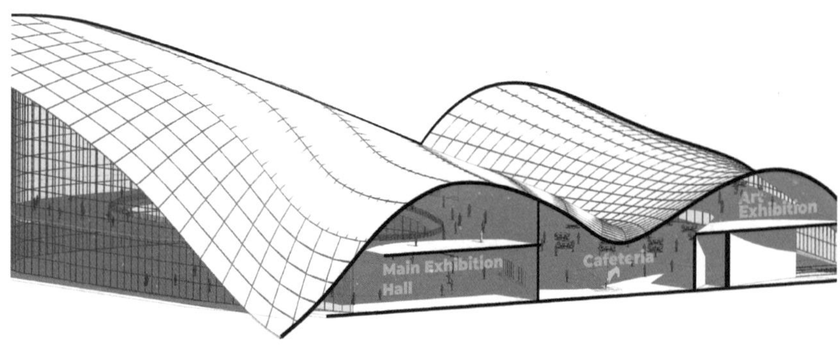

3D效果图 3D Shot

3D效果图 3D Shot

3.文化中心 Cultural Center

总体规划 Masterplan

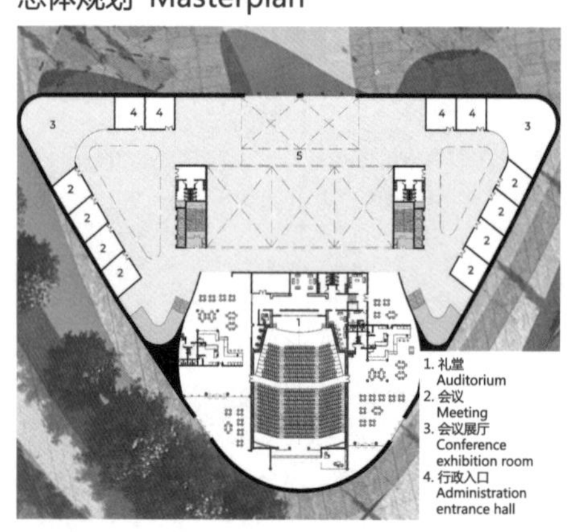

1. 礼堂 Auditorium
2. 会议 Meeting
3. 会议展厅 Conference exhibition room
4. 行政入口 Administration entrance hall

3D效果图 3D Shot

4.观景塔 Observation Tower

总体规划 Masterplan

1.机械室 Mechanical room
2.储物 Storage
3.电梯 Elevator

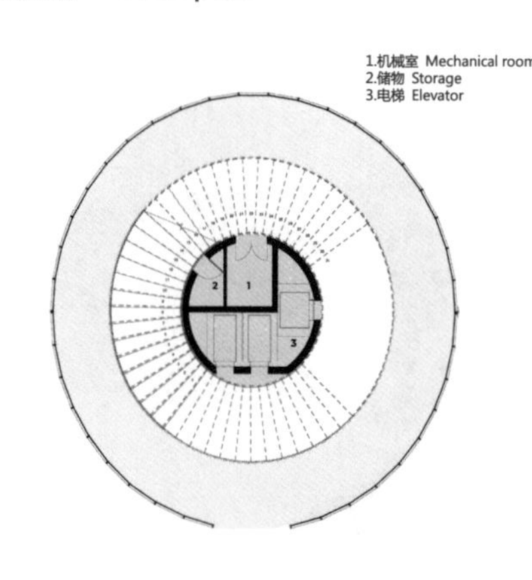

剖立面图 Sectional Elevations

课程感受 Course Feeling

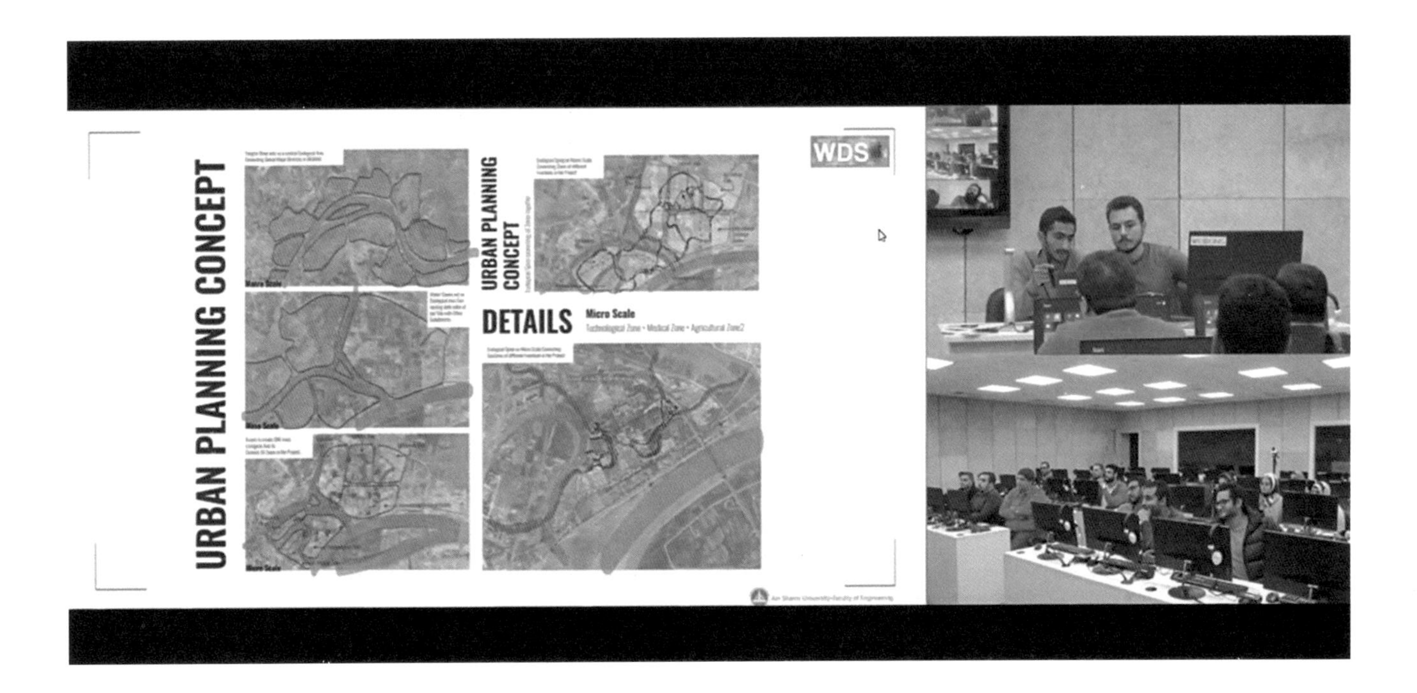

由于新冠肺炎的蔓延，开学被迫延期，本次课程采用了线上授课这种特别的方式，但遥远的空间距离没有阻碍大家热烈探讨，整个线上交流中充满着思想能量的交会与碰撞。在这个过程中，不仅设计课程得到持续深入与完善，同学们也越来越理解城市设计，培养了设计思维。

Due to the Covid-19, the start of school was forced to be postponed. This course uses a special method of online teaching, but the long distance cannot hinder everyone's enthusiastic discussion. The entire online communication is full of ideological energy encounters and collisions. This is not only the continuous deepening and improvement of the design works of each group, but also the deepening and improvement of the students' understanding of urban design and design thinking.

本次课程有多专业方向的老师把控，我们能够听到各方面的建议以更为全面地完善自己的思路，与美国和埃及的老师和同学的定期汇报交流更是打开了我们的视野，看到了在不同文化背景和思想领域催生的思维方式和设计方法，给予了我们新的启发。

This course is controlled by teachers from multiple disciplines. We can get suggestions from all aspects to improve our thinking more comprehensively. Regular reports and exchanges with teachers and students in the United States and Egypt have opened our horizons. Different cultural backgrounds and ideological fields have given birth to new ways of thinking and design methods that give us new inspiration.

对于风景园林专业的同学而言，城市设计是比较陌生的领域，在本次课程学习中，我们得以从更加宏观的角度看待城市发展的规律，并了解城市的经济、布局、交通等设施和要素与它们彼此之间的联系及影响，也学会了利用成熟的设计理论和信息技术去辅助和支持我们形成设计方案。在课程中也遇到了很多难题和困惑。一是对城市设计相关理念和实践的认知并不具体，对城市设计如何衔接上位规划、对接详细设计，以及其设计的深度与广度难以把握。二是风景园林设计逻辑向城市设计的视角和思维转换存在一定困难，以致限制了思维。从前期分析到场地的定位特色和设计概念的确定，其中的逻辑关系梳理也一度让人觉得十分头疼。三是在将概念、想法落实到具体的设计方面，同学们也遭遇了很多困难，但也终于在老师们的帮助下一点点体会如何用城市的布局、结构和空间从意象和功能等各方面体现设计主题，一步步完成从概念到总体设计的反复推敲与修正。

For students majoring in landscape architecture, urban design is a relatively unfamiliar field. In this course, we can look at the law of urban development from a more macro perspective, and understand the city's economy, layout, transportation, other facilities and also the relationship and influence between them. At the same time, we have learned to use mature design theory and information technology to assist and support the formation of design schemes as well. In this course, we also encountered many problems and confusions. Firstly, the cognition of related concepts and practices of urban design is not specific, and it is difficult to grasp how urban design connects with upper-level planning and detailed design, as well as the depth and breadth of its design. Secondly, there are certain difficulties in the transition from landscape design logic to the perspective and thinking of urban design, which restricts thinking. From the preliminary analysis to the determination of the location characteristics and design concepts of the site, the sorting out of the logical relationships also once made us feel very troublesome. Thirdly, how to implement the concept into the space and let the layout, structure and space of the entire city reflect its theme is also a place where students generally find it difficult in the design stage. At this stage, many groups have spent more class hours and finally, with the help and guidance of the teachers, the repeated deliberation and correction from the concept to the overall design are completed step by step.

依赖于网络的交流形式加大了沟通和合作的难度，同学们在线上讨论期间也会有分歧，但所有的争论都是为了做出最理想的设计，同学们也在不断的交流中加深了感情。非常感谢小组成员们积极的思想贡献和相互配合，以及对彼此的理解和包容，课程终于圆满完成，同学们也收获了真挚的友谊。因为新冠疫情的影响没有和外籍师生当面交流，也为这次课程留下一点遗憾，外籍老师的点评以及外校学生的汇报总会给我们带来不一样的启发，思考方式的差异让人看到解决问题的更多可能性，希望以后可以有和所有老师与同学一起面对面交流的机会。

The form of communication that relies on the internet has increased the difficulty of communication and cooperation. The students have disagreements and quarrels during online discussions. Every bit is to achieve the most ideal design. The students also deepen their feelings through continuous communication. It is very grateful for the group members' positive ideological contributions and mutual cooperation, as well as their understanding and tolerance of each other. Finally, the course was successfully completed and the students also gained sincere friendship. Because of the impact of the Covid-19, we have not been able to communicate face-to-face with foreign teachers and students, leaving a little regret for this course. The comments of foreign teachers and the reports of students from other schools will always give us different inspirations, and the differences in the way of thinking make us feel different and see more possibilities to solve the problem. I hope to have the opportunity to communicate face-to-face with all teachers and students in the future.

最后，真诚地感谢老师们这一个学期对我们的指导、帮助，参与这次课程使我们收获良多。感谢各位老师的辛苦付出，祝愿联合设计课程越来越好，未来的学弟学妹也能在此收获一份属于自己的成长与感悟！

Finally, I would like to express my sincere thanks to all the teachers for their guidance and help in this semester. We have gained a lot from participating in this course. Thank you for your hard work. I wish the joint design course is getting better and better, and future schoolmates can also gain their own growth and perception here!